MODEL BASED LEARNING AND INSTRUCTION IN SCIENCE

Models and Modeling in Science Education

Volume 2

Series Editor

Professor J.K. Gilbert
Institute of Education, The University of Reading, UK

Editorial Board

Professor D.F. Treagust
Science and Mathematics Education Centre, Curtin University of Technology,
Australia
Assoc. Professor J.H. van Driel
ICLON, University of Leiden, The Netherlands
Dr. Rosária Justi
Department of Chemistry, University of Minas Gerais, Brazil
Dr. Janice Gobert
The Concord Consortium, USA

For other titles published in this series, go to
http://www.springer.com/springer_series_in_computer_science

Model Based Learning and Instruction in Science

John J. Clement
Editor

University of Massachusetts, Amherst, USA

Mary Anne Rea-Ramirez
Editor

Western Governors University, Salt Lake City, USA

Editors

John J. Clement
University of Massachusetts
Amherst
USA

Mary Anne Rea-Ramirez
Western Governors University
Salt Lake City
USA

ISBN 978-1-4020-6493-7   e-ISBN 978-1-4020-6494-4

Library of Congress Control Number: 2007938402

© 2008 Springer Science + Business Media B.V.
No part of this work may be reproduced, stored in a retrieval system, or transmitted
in any form or by any means, electronic, mechanical, photocopying, microfilming, recording
or otherwise, without written permission from the Publisher, with the exception
of any material supplied specifically for the purpose of being entered
and executed on a computer system, for exclusive use by the purchaser of the work.

Printed on acid-free paper.

9 8 7 6 5 4 3 2 1

springer.com

Contents

Contributors vii

Acknowledgement ix

Model Based Learning and Instruction in Science 1
   John J. Clement

Section I: Basic Concepts and Background for Model Based Learning

1. Student/Teacher Co-construction of Visualizable Models
   in Large Group Discussion 11
   John J. Clement

2. An Instructional Model Derived from Model Construction
   and Criticism Theory 23
   Mary Anne Rea-Ramirez, John Clement and Maria C. Núñez-Oviedo

Section II: Introduction to Model Based Teaching Strategies

3. Determining Target Models and Effective Learning Pathways
   for Developing Understanding of Biological Topics 45
   Mary Anne Rea-Ramirez

4. Co-construction and Model Evolution in Chemistry 59
   Samia Khan

5. Target Model Sequence and Critical Learning Pathway
   for an Electricity Curriculum Based on Model Evolution 79
   Melvin S. Steinberg

6. Case Study of Model Evolution in Electricity: Learning
   from Both Observations and Analogies 103
   John Clement and Melvin S. Steinberg

Section III: Qualitative Research on Specific Strategies

7. A Competition Strategy and Other Modes for Developing Mental Models in Large Group Discussion 117
   Maria C. Núñez-Oviedo and John Clement

8. What If Scenarios for Testing Student Models in Chemistry 139
   Samia Khan

9. Applying Modeling Theory to Curriculum Development: From Electric Circuits to Electromagnetic Fields 151
   Melvin S. Steinberg

10. Developing Complex Mental Models in Biology Through Model Evolution 173
    M.C. Núñez-Oviedo, John Clement and Mary Anne Rea-Ramirez

11. Role of Discrepant Questioning Leading to Model Element Modification 195
    Mary Anne Rea-Ramirez and Maria Cecilia Núñez-Oviedo

12. Using Analogies in Science Teaching and Curriculum Design: Some Guidelines 215
    Mary Jane Else, John Clement and Mary Anne Rea-Ramirez

13. Model Based Reasoning Among Inner City Middle School Students 233
    Mary Anne Rea-Ramirez and Maria Cecilia Núñez-Oviedo

14. Six Levels of Organization for Curriculum Design and Teaching 255
    John Clement

Author Index 273

Subject Index 277

Contributors

Clement, John, Professor
    University of Massachusetts, Amherst
    813 No. Pleasant Street
    Amherst, MA USA 01003

Else, Mary Jane
    University of Massachusetts, Amherst
    813 No. Pleasant Street
    Amherst, MA USA 01003

Khan, Samia, Faculty of Education
    University of British Columbia
    2125 Main Mall
    Neville Scarfe Building
    Vancouver, BC
    CANADA V6T 1Z4

Núñez-Oviedo, Maria Cecilia, Professor
    Jefe de Carrera de Pedagogia en Cs. Nat. y Qca.
    Of 317, Facultad de Educación
    Universidad de Conception
    Edmundo Larenas 335
    Concepcion, Chile

Rea-Ramirez, Mary Anne, Professor
    Western Governors University
    Teachers College
    4001 South 700 East Suite 700
    Salt Lake City, UT USA 84107
    mramirez@wgu.edu
    877-435-7948 ext. 1899
    801-274-3280 ext. 1899
    804-339-8117 (cell)

Steinberg, Melvin, Professor (emeritus)
    Smith College
    McConnell Hall 410
    Northampton, MA USA 01003

Acknowledgement

The material in this book is based upon work supported by the U.S. National Science Foundation under Grants ESI-9911401, REC-0231808, and DRL-0723709. Any opinions, findings, and conclusions or recommendations expressed are those of the author(s) and do not necessarily reflect the views of the National Science Foundation.

Model Based Learning and Instruction in Science

John J. Clement
University of Massachusetts, Amherst

Introduction

This book describes new, model based teaching methods in science instruction and presents research results on their characteristics and effectiveness. It includes discussions of teaching methods based on evidence from a diverse group of settings: middle school biology, high school physics, and college chemistry classrooms. Mental models in these areas such as understanding the structure of the lungs or cells, molecular structures and reaction mechanisms in chemistry, or causes of current flow in electricity are notoriously difficult for many students to learn. Yet these models lie at the core of conceptual understanding in these areas.

Each of the classrooms studied by the authors used recently developed curricula that fostered unusually active learning processes. The curricula were designed to create flexible mental models in students as a key source of understanding. The studies focus on a variety of teaching strategies such as discrepant questioning, analogies, animations, model competition, and hands on experiments. Through in depth case studies of curricula and actual classroom transcripts, they attempt to unpack and analyze the nature of the teaching strategies being used at a number of levels.

A pressing need is to address the problem that many of today's teachers feel pulled in two different directions: on the one hand they are urged to teach content in a broader and deeper way as measured by standardized tests. On the other hand, they are urged to adopt student centered inquiry methods. Teachers often feel that these goals are incompatible. The strategies described in this book are midway between open discovery and lecture approaches, and provide important models for teachers facing this dilemma. The chapters focus on new instructional strategies for attaining deeper levels of conceptual understanding in science and develop new diagrammatic representation systems for analyzing these strategies. Six different levels of organization for teaching strategies are described, from those operating over months (design of the sequence of units in a curriculum) to those operating over seconds (teaching tactics for guiding discussion). Strategies at different levels are diagrammed in different ways. Combined with concrete examples of the strategies, drawn from real curricula and classroom teaching exchanges, this provides a way to understand more clearly each strategy and the overall approach to model based teaching and learning.

This book is divided into three major sections. Section I introduces chapters on the basic concepts and background of model based learning. We suggest a model of mental model co-construction that is central to our understanding about how students construct understanding in science. We also look briefly at the literature on classic conceptual change, social learning theory, mental model theory, and the contribution of history of science literature on the field.

Section II investigates the concepts of determining target models and co-construction in biology, chemistry, and physics domains. While we present some data in these chapters, they are primarily intended to introduce the concepts and strategies of model based learning. Following this, Section III presents qualitative studies of strategies used in model based teaching and learning. Finally, we conclude with a discussion of the interconnection of strategies among the different studies and the different areas of science. The following provides a brief overview of each chapter.

Section I: Basic Concepts and Background for Model Based Learning

Chapter 1: Student/Teacher Co-construction of Visualizable Models in Large Group Discussion

The approaches to teaching described in this book use a common strategy of student-teacher co-construction that builds on student-generated as well as teacher-generated model elements. This strategy, and some of the issues it raises, is introduced in this chapter. The primary example used is a markedly different approach to teaching biology at the middle school level. It comes from an experimental eighth grade curriculum titled "Energy and the Human

Body" that deals with pulmonary and cellular respiration, circulation, and digestion (Rea-Ramirez, Clement, Nunez, and Else, 2004). Using examples from student discussion, basic concepts used in the rest of the book –"co-construction", "scaffolding", "reasoning zone", "agenda setting" and "model evolution"– are introduced. Diagramming techniques are used to give the reader a picture of these special types of teacher-student interaction.

Chapter 2: An Instructional Model Derived from Model Construction and Criticism Theory

In this chapter we present an overview of the theories that led up to the current theory of model based teaching and learning that underlies this book, along with an introduction to model-based co-construction. In particular, we recognize the important role of conceptual change and social learning theories in previous research and curriculum development efforts. However, while these theories have contributed to progress in the field, each has had limitations in describing the deep cognitive activity that occurs as a student and teacher co-construct understanding. In order to better situate the research presented in this book, a brief description of each of these prior theories is included. This is not meant to be a full review of the areas but rather an introduction that provides a base for the theories of model based learning and co-construction of knowledge. The present book concentrates on an analysis of specific teaching strategies at different time scale levels. It combines the cognitive analysis of conceptual change processes in students with the examination of instructional dialogue in which socio-cognitive methods are used to encourage students to build scientific models. This can be seen as a contribution to the efforts to integrate social and cognitive perspectives for explaining science instruction. Thus the review frames our work as contributing to an integration of cognitive and social perspectives on learning.

Section II: Introduction to Model Based Teaching Strategies

Chapter 3: Determining Target Models and Effective Learning Pathway for Developing Understanding of Biological Topics

Prior research has indicated that students of all ages show little understanding of respiration beyond breathing in and out and the need to breathe for survival. This occurs even after instruction, with alternative conceptions persisting into adulthood. Whether this is due to faulty educational strategies or to the level of difficulty in understanding a complex system is an important question. When expectations of learning are developed from the expert's view, it may be that we are setting students up for failure. A critical look at how much and what type of knowledge is necessary to construct a worthwhile understanding of complex topics is an important first step. This entails determining an integrated target model and an effective learning pathway that provides realistic chances for real understanding. The proposed pathway used as an example in this chapter is aimed at promoting integrated learning that is deep enough to enable middle school students to explain the path of oxygen molecules to the lungs and glucose from food into the bloodstream, cells, and their eventual role in the mitochondria.

Chapter 4: Co-construction and Model Evolution in Chemistry

Co-construction and model evolution are described in this chapter as processes teachers use to guide students to enrich their original models so that they have greater explanatory and predictive power. An instructional interaction between a teacher and a pair of students is presented. This learning episode illustrates how a chemistry teacher co-constructed a model of molecules with students that could explain vaporization and boiling. The teacher first encouraged students to express their model of molecules by focusing on single relationships within a model one at a time. Students constructed relationships and successively added variables to their models, resulting in models that had evolved and were enriched compared with what they expressed earlier in an initial survey. The teacher then asked the students to test their model. Students' ideas about molecular structures appeared to evolve specifically when students were asked to run a test. This interaction appeared to motivate students to consider new factors that they had not expressed before. The episode reveals how students were able to express, enrich, and evolve their models in chemistry with questions and activities from their teacher in a co-constructive process.

Chapter 5: Target Model Sequence and Critical Learning Pathway for an Electricity Curriculum Based on Model Evolution

This chapter discusses the way principles of qualitative modeling are applied to the design of a high school electricity curriculum called the

Capacitor-Aided System for Teaching and Learning Electricity (CASTLE). The curriculum aspires to enable students to build a sequence of increasingly complex models of current propulsion in circuits. Model building is fostered by hands-on experiments with bulb lighting in circuits that contain batteries and capacitors. These are sequenced to foster a learning pathway of incremental model modifications that add complexity with low cognitive load. The sequence producing changes in the student's model of a current-driving agent from emission by a battery to pressure in a compressible fluid is described. Model modification is stimulated by surprising observations that require the student to modify the model to account for new circuit behaviors, some of which are strongly at odds with deeply held preconceptions. The curriculum is designed to foster model modification by transfer of dynamic imagery from concrete analog situations to the model. The goal is to enable students to use imagistic simulations to evaluate proposed model modifications and to solve novel problems.

Chapter 6: Case Study of Model Evolution in Electricity: Learning from Both Observations and Analogies

This chapter focuses on a tutoring case study in high school level instruction using the curriculum on electric circuits described in the previous chapter. A model evolution approach to instruction is described as working within a cycle of model generation, evaluation, and revision. Transcripts from the lessons allow one to develop models in the form of diagrammatic representations of the learning and teaching processes involved. Basic strategies described include 1) Model Evolution: Instead of a "remove and replace" view of learning, the approach uses a "transformation" view that modifies aspects of an existing intermediate model; 2) Discrepant Events: Discrepant events constrain the construction of the new model as well as creating dissonance with the old model. The approach involves a series of discrepant events rather than relying on one discrepant event; and 3) Analogies approach uses a series of analogies for adding pieces to a model over an extended period of model construction, rather than relying on one quick analogy.

In this approach, observations from experiments are used often during the interaction but are not viewed as sufficient for producing normative conceptual change. Positive sources of knowledge from students' experiential memories are also used. The process relies on using both rational-analogical (prior knowledge) as well as observational sources of ideas. Diagrams of the learning process reflect the theoretical position that contributions to the student's evolving model are made both "from above" and "from below" as observations interact with prior conceptions. Thus, the process fits an interactionist view of learning as empirically constrained, creative model construction rather than a view of learning as generalizing from observations.

Section III: Qualitative Research on Specific Strategies

Chapter 7: A Competition Strategy and other Modes for Developing Mental Models in Large Group Discussion

This chapter describes a teaching strategy for guiding large group discussions called Model Competition. The teacher has an opportunity to promote Model Competition when the students contribute to a discussion with ideas that are contradictory to each other. The presence of these different kinds of ideas can foster dissatisfaction and curiosity in the students' minds that can be productive. The strategies a teacher uses to support this in a case study of classroom learning in the area of respiration are examined. The teacher played a key role during the teacher/student co-construction process by constantly diagnosing the students' ideas and encouraging the students to disconfirm, recombine, restructure, or tune their ideas and to generate successive intermediate mental models.

Chapter 8: What If Scenarios for Testing Student Models in Chemistry

This chapter describes a strategy aimed to help students test their mental models in chemistry. The strategy involves encouraging students to create 'what if' scenarios. 'What if' scenarios involve speculating on and changing one or more of the parameters of a model and observing the effects of the change. 'What if' scenarios can be used after a student has expressed an initial model of a phenomenon. In one example, regarding their model of charged particles, the teacher asked students a series of what if questions that changed a single variable, such as magnitude of charge, in increments. The students' models of charged particles appeared to transform as they explored the boundaries of their model in increments. Thus 'what if' scenarios are a teaching strategy that has the potential to enrich or transform students' existing models through testing. There are a variety of methods that aid in supporting what if scenarios, and this chapter offers several examples from the field of chemistry.

Chapter 9: Applying Modeling Theory to Curriculum Development: From Electric Circuits to Electromagnetic Fields

This chapter describes the efforts the designers of a high school electricity curriculum made to enable students to construct increasingly complex models of electrostatic and then electromagnetic distant action. The electrostatics sequence begins where the target model sequence ended in Steinberg's earlier chapter on modeling current propulsion in circuits. It describes low-tech means for investigating phenomena that can generate imagery of "potential pressure halos" in the space around charge that act on other charge located elsewhere. (Halo diagrams are visual representations of electric potential functions, and "potential pressure" is intuitive language for "electric potential".) Lighting by neon bulbs identifies negative mobile charge carriers ("electrons") which are tiny compared to atoms and are present but bound in insulator atoms. The electromagnetic sequence describes how coaxial coils, LEDs and AM radio provide evidence that curling electric field is radiated by accelerating charge during current turn-on and turn-off and is accompanied by curling magnetic field, explaining magnetic field production and transformer action. The progression of diagrams used in the curriculum represents a careful building up of a qualitative model that is designed to enable students to make sense of the vector electric field concept. This progression again embodies the principles of Gradual Model Modification – whereby sense-making continuity and model runnability are maintained via small-step model revisions.

Chapter 10: Developing Complex Mental Models in Biology Through Model Evolution

This chapter describes a second study of middle school classes learning about human respiration. It addresses the questions: "What processes are involved in the co-construction of mental models?" and "What are the teacher's and the students' contributions to the co-construction process?" To answer these questions the authors worked from patterns in transcripts of classroom dialogues to produce diagrammatic models of the interactions and different teaching strategies used. A central theme in the instructional approach is learning as model evolution. The overall model developed is a theory that includes several layers of nested teacher-student interaction patterns that can be observed during instruction. This study also describes how the teacher combined these large, intermediate, and small sized interaction patterns to build successive intermediate mental models with her students. By developing a model of the cognitive effects of this social process, we describe how this co-construction process lies at an intermediate point between didactic approaches and discovery learning approaches.

Chapter 11: Role of Discrepant Questioning Leading to Model Element Modification

This chapter examines one strategy for promoting model-based co-construction called discrepant questioning. It has been well established that questioning is an important aspect of good teaching. However, the effect that specific questioning strategies have on cognitive processes that stimulate mental model construction is less well researched. Cases are examined suggesting that student-student and student-teacher interaction based on and following discrepant questioning appears to stimulate dissatisfaction leading to model element modification. This in turn should aid students in developing more dynamic, integrated mental models. Examples of student generated responses and model element generation and modification are presented.

Chapter 12: Using Analogies in Science Teaching and Curriculum Design: Some Guidelines

In this chapter, we present guidelines for the use of analogies in science teaching that are based both on research in the science education literature and on three years of our own classroom curriculum-trial experiences. We consider analogies to be a key tool in science teaching because analogies serve as preliminary models that are familiar to students. These preliminary models are then built upon through the use of other methods such as demonstrations and teacher-student discussions. This process is student-active in that students are asked to explore the analogy and subject it to reasoning checks that help them understand the ways in which it is similar to and dissimilar from the knowledge "target." Using analogies in this active way may have the added benefit of helping students become more aware of their evolving mental models. Our observations suggest that successful analogy use depends both upon this student involvement and upon careful guidance by teachers. We suggest that teachers and students will profit from using analogies in science teaching when they are aware of the both the strengths and possible pitfalls of analogies. We explore these pitfalls, classify analogies by goals and complexity, and present examples of successful and unsuccessful analogy use.

Chapter 13: Model Based Reasoning Among Inner City Middle School Students

This chapter describes an experimental curriculum trial in an inner city setting using the Energy in the Human Body curriculum described in previous chapters. Although these students came from traditionally under represented groups, they showed significant pre-post change on tests of conceptual understanding and gave evidence for model based reasoning. Most students' scores indicated that they had achieved the target model. Several of the other students showed improvement and only one student showed no change. We believe that separating the dimensions of topic, structure, function, dynamic modeling, causal chains, depth of understanding, and integration across topics in the analyses of student responses gives us a much more fine grained measure of student understanding and a richer perspective on what students are learning.

Chapter 14: Six Levels of Organization for Curriculum Design and Teaching

Six levels of organization for teaching strategies are described as a way of summarizing the findings in the book across the three curricula and age levels examined. The levels range from those operating over months (design of units in a curriculum) to those operating over seconds (teaching tactics for guiding individual responses in discussion). Chapter findings are placed at the different levels yielding the outline of an overall theory of instruction based on student-teacher co-construction of mental models. The chapter concludes that instructional design should include all six levels to be optimally effective in teaching for meaningful conceptual change leading to integrated knowledge that can be applied flexibly.

Section I

Basic Concepts and Background for Model Based Learning

Chapter 1
Student/Teacher Co-construction of Visualizable Models in Large Group Discussion

John J. Clement

University of Massachusetts, Amherst

1.1 Introduction

One of the goals of this book is to address some of the conflicts teachers feel in responding to the calls for both inquiry and conceptual understanding content goals. Teachers often feel that open-ended methods are incompatible with strong content goals that they are asked to fulfill. The central strategy used in the approaches described in this book takes an intermediate position. The strategy is one of student-teacher co-construction that elicits student generated model elements as well contributing some from the teacher (Rea-Ramirez, 1999; Steinberg & Clement, 2001; Nunez, Ramirez, Clement, & Else, 2002; Clement & Steinberg, 2002). I will introduce this strategy and some of the issues it raises in this chapter.

1.2 Content Goals and Target Models

The example of a lesson sequence that I will use in this chapter comes from an experimental 8th grade curriculum titled "Energy and the Human Body" that deals with pulmonary and cellular respiration, circulation, and digestion (Ramirez, Nunez, Clement, and Else, 2004). Content goals are taken seriously in this curriculum and are expressed as target models (also referred to by John Gilbert and others as "curriculum models" [Buckley, Boulter, & Gilbert, 1997]). Each target model is a desired knowledge state that one wishes students to posses after instruction, such as an image of a lung filled with alveoli, branching air tubes, and adjacent branching capillary structures. These may not be as sophisticated as the expert consensus model currently accepted by scientists. Instead of logical relationships used in formal treatments of the topic, an educator's view of the target model reflects qualitative, simplified, analogue, and tacit knowledge that is often not recognized by experts.

The curriculum must deal effectively with the problem that content goals in science are sometimes frustrated by the presence of student misconceptions (sometimes termed alternative conceptions to remind us that we need to respect them as student constructions that are often reasonable ideas) which are in conflict with the target model. However we also are alert for student conceptions that are compatible with current scientific models and that can be used as building blocks for developing the target model (Clement, Brown, & Zietsman, 1989).

Van Zee and Minstrell (1997a, 1997b) have discussed a number of strategies for promoting large group whole class discussion by drawing out students' ideas within the context of teaching for conceptual change in the presence of alternative conceptions. Hammer (1995) has documented some impressive thinking processes that can occur under such conditions in a secondary physics course. Further work is needed, however, on precisely how large group discussions can feed cognitive model construction processes that are aimed toward content goals. Here I want to examine ways to describe some of the roles teachers can play when they allow student ideas, both correct and incorrect, to be taken seriously in classroom discussions (where by "correct" I mean largely compatible with the target model for the lesson).

1.3 Instructional Approach used

The topics covered in the Energy and the Human Body curriculum include digestion, pulmonary respiration, the distribution of oxygen and sugar by the circulatory system, and microscopic respiration in the mitochondria. The instructional strategies used include hands-on activities, analogies, discrepant events, model building, and computer generated animations, supported by scaffolding and probing questions. In this chapter I will focus on a short example of the approach used in the *pulmonary respiration sequence* to illustrate some of the discussion leading decisions faced by the teacher.

A principle teaching method used to aid mental model construction is to have students invent models of body systems that could perform functions like breathing or delivery of nutrients to a limb. They are asked to do this on their own initially before receiving information from the teacher. Almost always this involves making drawings. The teacher then uses the students' initial models (including misconceptions contained therein) as a starting point to foster a series of model criticisms and improvements. Eventually enough changes are made to approach the target model for the lesson.

1.3.1 Model Evolution

Figure 1.1 shows a typical part of the lungs teaching interaction. In initial models constructed in a drawing individually by each student, the lung is mostly hollow (a partially incorrect "balloon" model of the lungs). A student then draws a collective model at the board in front of the group, with veins and many hair-like structures on the interior surface to "filter bad stuff out of the air." There is also a hole at the bottom of the lung. The teacher then asks a "discrepant question", Does your model show air going out the bottom of the lung into the body? The students then begin to worry about air leaving the lung there and decide to modify their model by closing the hole. This is an example of an indirect and mild but focused intervention by the teacher. Some students think that the two parts of the lung might actually be joined together to in effect form one large cavity. At that point the teacher asks another discrepant question: "Are there operations where they remove one lung?" Students agree they know of such operations and decide that there must be two separate lungs.

Later when students realize the lungs are more "meat like" than "balloon like," they make the air passageways too few in number to hold enough air. At that point the teacher sets up a breath measurement experiment where they use long plastic bags to measure how much air is contained in a deep breath, showing that there must be many pervasive tubes and passages indeed. Thus the discussion led by the teacher modifies the model in small steps, making it more and more like the target model for the lesson (Clement & Steinberg, 2002).

In describing this process three aspects of Fig. 1.1 are noteworthy:

- *An Evolutionary Sequence of Models and Revisions.* The sophistication of the students' explanations grew steadily during the instructional treatments. We can view the students' conceptual changes here as producing a sequence of intermediate models that become progressively more expert-like. This suggests an overall view of learning that has *model evolution* as its central feature, where students are able to build on knowledge that they had developed in earlier sections (Steinberg & Clement, 1997, 2001; Clement, 2000). This evolution is depicted in Fig. 1.1 as the sequence of models that change from left to right in the middle row of the figure.
- *Starting from the Students' Ideas.* The first model that the students draw comes before virtually any instruction on the lung system. Therefore it expresses their prior knowledge and initial creative ideas about the structure of the lungs. This includes correct, incorrect, and partially correct features. The approach taken here shows that these partially correct pre-conceptions can be modified and built upon. This amounts to a change from a "remove all prior knowledge and replace it" view of conceptual change and learning to a "gradual transformation" view that works with and modifies aspects of the student's existing model.
- *Discrepant Questions Or Events.* These were used to motivate model revisions. These included the breathing capacity measurement. We modeled effects of the discrepant questions as producing internal dissonance with the current model. These dissonance relations are shown as jagged lines in Fig. 1.1.

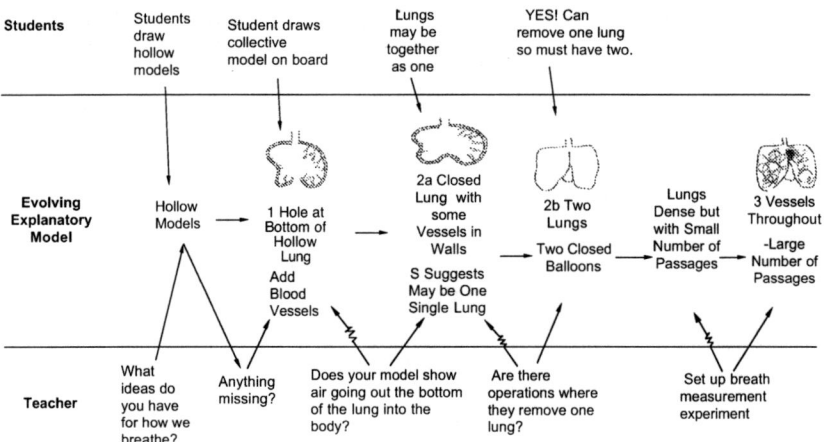

Fig. 1.1 Co-construction in pulmonary system

Model-based learning and instruction

J. Clement

Q181.M6 2008 eb

Model-based learning and instruction

J. Clement

0.18, MB p3826

1.3.2 Distinguishing Between a Student Directed Agenda and Student Generated Ideas

We believe that part of the dilemma faced by teachers with both content goals and calls to use student centered inquiry strategies may be solved by adopting guided inquiry techniques. However, to describe these techniques requires us to increase the precision of the vocabulary that we use to describe classroom interactions.

Two important but different ways to talk about student centeredness in a curriculum are the extent to which:

- Activities are teacher or student *directed* (Who is setting the questions and the agenda?)
- Ideas are teacher or student *generated* (Who is generating and evaluating the explanations and ideas in the learning?)

These are separate dimensions for describing a classroom but they are often confused. A way of describing the intent of the present curriculum is that it is teacher directed about 85% of the time–the teacher carefully directs the attention of the students to most topics and activities in a planned sequence. Thus it is quite teacher directed. Yet its ideas are teacher generated directly only about 40% of the time: within each topic students are encouraged to propose as many ideas as possible and then to modify and improve them, so that they may end up proposing 60% or more of the ideas. Thus the knowledge developed is largely student generated but at the same time the agenda is largely teacher directed.

This is a bit like the efficient structure of a meeting for an organization that has a chairman but that needs strong input from its members. The chairman sticks faithfully to the agenda for the meeting, but opens the floor for input on each agenda item. Creative responses are encouraged. In addition the chair draws out or reminds the members of constraints that force reconsideration or modifications in some of the ideas that come up. This structure combines openness to ideas with the efficiency of an agenda that allows one to achieve goals and prevents too much wandering from the topic. The structure contrasts to more dictatorial ones in that the members feel an investment in the outcome in that they have had an input to the process.

This puts the approach midway between pure "lecture" and pure "discovery learning", where by the latter I mean students inventing all the ideas without teacher input. The approach represents an intermediate position on whether ideas in classroom discussions should be teacher generated or student generated since it advocates both sources as important. In general, the aim is to have as many of the ideas be student generated as is practical, given the constraints of limited time for each curriculum topic. This serves the larger goal of fostering active learning and reasoning as a way to increase sense making, comprehension and retention. To do this the curriculum provides some guidance as to which ideas the students may be able to construct and which ideas usually require teacher introduction. Throughout discussions the teacher monitors the students' ideas, offering mild or if necessary, stronger support tactics to promote student construction until the next targeted model is reached.

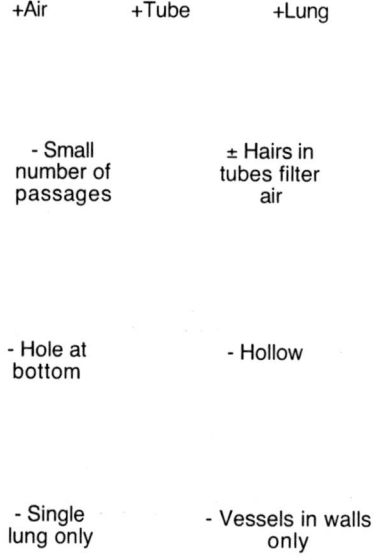

Fig. 1.2 Mosaic of student generated ideas for structure of lungs

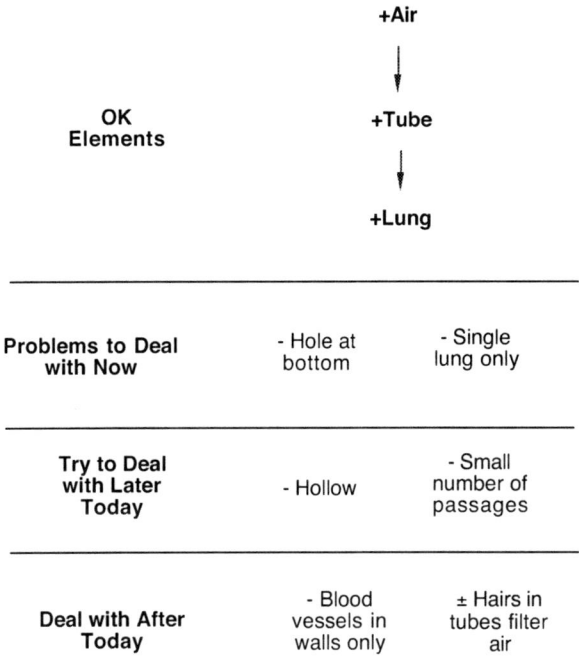

Fig. 1.3 Teacher's agenda organization of mosaic of ideas for structure of lungs

1.3.3 The Mosaic of Student Ideas Generated by Large Group Discussions

Once students begin to accept the invitation to contribute ideas to modeling questions, this poses new challenges to the teacher. A collection of unnervingly diverse ideas, both correct and incorrect, can be offered by students. The diagram in Fig. 1.2 shows an example of what Maria Nunez calls a "mosaic" of student ideas that the teacher is dealing with at any given moment. The mosaic outlines the collection of ideas that have been introduced, mostly by the students, in the large group discussion. Some of the ideas are largely correct in the sense of being close to the target model. Others are largely incorrect, and still others are partly correct.

To help steer their decision making within this somewhat complicated mix, teachers need to devise an agenda that reflects what they think should be dealt with first, second, etc. A teacher may reorganize the students' ideas into an agenda like that depicted in Fig. 1.3, raising the following additional issues: There may be natural connections between the largely correct ideas according to biological structures or functions (as indicated by the arrows). These can form a rough initial student model from which to work. The teacher will want to make sure these initially correct features are included in a drawing in front of the class. Teachers can form an agenda by sorting the largely incorrect and partially incorrect ideas into three categories: those we can work on now, work on later today, or work on after today. Rather than trying immediately to replace each misconception as it arises, the present teacher recognized that some of the misconceptions were too complex to deal with effectively immediately. Instead, she tried to work *on one difficulty at a time*, either by helping students to modify some aspect of their conception or replace it. She started with the most basic misconceptions, which once revised, would help her deal with more complex misconceptions. The last "Deal with After Today" category implies that teachers can postpone dealing with certain misconceptions until students are prepared to deal with them. This is usually counterintuitive for teachers however, and it requires practice.

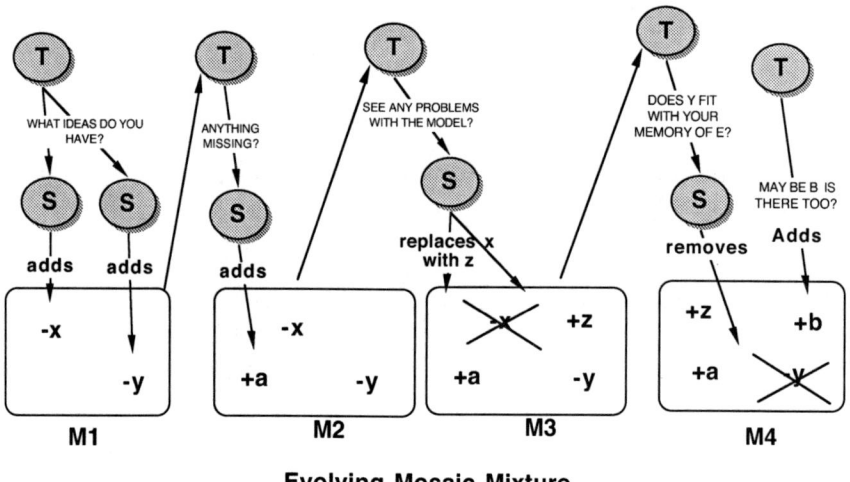

Evolving Mosaic Mixture

Fig. 1.4 Model evolution and co-construction in large group discussion

1.4 View of Large Group Model Construction in the Classroom

In this section I try to paint an image of large group discussion that may help teachers think about their role in guiding the process of student model construction. Figure 1.4 shows a more abstract version of the discussion process in order to highlight certain features (adapted from Clement, 2002). It represents an "evolving mosaic mixture" or "model evolution" pattern of large group discussion. Different students contribute correct and incorrect ideas during this evolution. Here a series of false to partially correct to more correct models are developed progressively, as was illustrated in Fig. 1.1. The teacher first draws out the students' ideas, both correct and incorrect. Discrepant questioning on specific issues by the teacher then triggers student generated corrections or additions to part of the model to eventually form intermediate model M4. This process continues to form more intermediate models until the target model is reached. Figure 1.4 shows this piecewise revision process as elements are changed one at a time. In this hypothetical example students are contributing about 80% of the ideas (as requested or prompted by the teacher) as opposed to the teacher's 20%, but in this curriculum the intended student contribution may vary on different topics from 25% to 90%. Figure 1.5 helps us clarify our vocabulary by showing how a term like "Model Evolution" is conceived of graphically. (Chapter 10 will present other modes of large group discussion such as a "model competition" pattern.)

An important aim of the teacher here is to keep students in a "Reasoning Zone". Building on Vygotsky's ideas, I define the Reasoning Zone as an area of discussion where students can reason about ideas and construct new ideas productively (or at least contribute to its production in a group). Not all student ideas move in the direction of the target, but if thinking in the Reasoning Zone includes idea *evaluation and modification*, then problems with false leads will be discovered and progress toward the target should occur. However, if the question or topic chosen by the teacher is too large or too hard, it will be outside of this Reasoning Zone. This is what makes it important to utilize a strategic agenda as illustrated in Fig. 1.3 to keep the students in a reasoning zone where they are able to make inferences and corrections to the growing model. Even so, sometimes discussion bogs down. In this case the teacher attempts to provide just enough support, or "scaffolding", in the form of a leading question, hint, new observation, reference to an earlier comment, discrepant question, or piece of information, etc. in order to get student reasoning going again. This is shown in Fig. 1.5 by the box labeled "scaffolding" to indicate that the teacher is supporting the students' reasoning.

1.5 Co-Construction

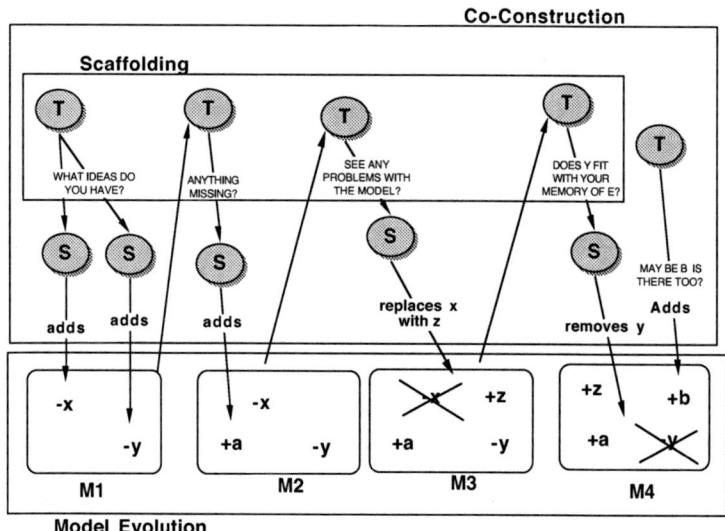

Fig. 1.5 Graphic illustrations of model evolution, scaffolding, and co-construction

If the students have some prior knowledge in the topic then many of the ideas can come from the students and not the teacher. In other cases some of the evaluations or modifications will be made by the teacher when specific content goals are a priority, a student modification cannot be elicited, and the teacher feels they are ready for it. Such patterns represent a process of *co-construction* in which both teacher and students contribute ideas and evaluations of ideas. This process is shown by the box under that name in Fig. 1.5.

The Energy in the Human Body curriculum is rich in visualizable models, therefore we believe it is important for the teacher to help students communicate with each other by drawing what students are describing (whether correct or not) on the board in front of everyone or having students draw their own models on the board or on posters or whiteboards in small group work. This provides a visual as well as a verbal communication channel to foster discussion. Drawings can then be modified to reflect modifications as the discussion proceeds. The value of such drawings for focusing the discussion on the evolving model cannot be understated.

1.6 Summary

If one starts a unit by drawing out the students' ideas, one taps into both correct and incorrect ideas. However these ideas can be modified and built upon gradually through cycles of evaluation and revision. This leads to the class generating an evolutionary sequence of intermediate models that can move close to the target model. However, allowing students to generate ideas can allow several misconceptions at once into the discussion. Handling this requires the teacher to form an agenda in order to guide discussions to easier issues first so that these can be dealt with in the students' "reasoning zone". It also requires scaffolding students' reasoning via techniques such as guided discussion and "discrepant questions". We refer to such an interaction where ideas come from both the teacher and the students as *"teacher-student co-construction"*. The resulting pattern in Fig. 1.4 is unusual in the extent to which it uses student-generated ideas. Using this mode of discussion along with a mosaic agenda and discrepant questioning may be challenging to orchestrate at first, but it holds much promise as a pedagogical strategy that can foster both inquiry skills and content learning.

The remainder of this book examines strategies similar to this one in various domains and the conditions under which each one may or may not be successful. The chapters that follow describe different kinds of classroom interactions that involve a large number of student generated ideas, and recommend a variety of specific pedagogical strategies, in addition to discrepant questioning, that respect the students' power to generate and evaluate useful ideas but that also take seriously specific conceptual goals in the form of target models and the need for the teacher to support and contribute to the process.

References

Buckley, B., Boulter, C., & Gilbert, J. (1997). Towards a typology of models for science education. In J. Gilbert (Ed.), *Exploring models and modeling in science and technology education*, (pp. 90–105) Reading: The University of Reading.
Clement, J. (2000). Model based learning as a key research area for science education. *International Journal of Science Education, 22*(9), 1041–1053.

Clement, J. (2002). Managing student/teacher co- construction of visualizable models in large group discussion. *Proceedings of the AETS 2002 Conference*, Charlotte, N. C.

Clement, J., Brown, D., & Zietsman, A. (1989). Not all preconceptions are misconceptions: Finding anchoring conceptions for grounding instruction on students' intuitions. *International Journal of Science Education, 11,* 554–565.

Clement, J. & Steinberg, M. (2002). Step-wise evolution of models of electric circuits: A "learning-aloud" case study. *Journal of the Learning Sciences 11*(4), 389–452.

Hammer, D. (1995). Student inquiry in a physics class discussion. *Cognition and Instruction, 13,* 401–430.

Nunez, M., Ramirez, M., Clement, J., & Else, M. (2002). Teacher-student co-construction in middle school life science. *Proceedings of the AETS 2002 Conference.*

Rea-Ramirez, M. A. (1999). Explanatory need. *Proceedings of the National Association of Research in Science Teaching Meeting,* Boston, MA.

Rea-Ramirez, M. A., Nunez-Oviedo, M. C., Clement, J., & Else, M. J. (2004). *Energy in the Human Body Curriculum.* University of Massachusetts, Amherst.

Steinberg, M., & Clement, J. (1997). Constructive model evolution in the study of electric circuits. *Proceedings of the International Conference 'From Misconceptions to Constructed Understanding',* Cornell University.

Steinberg, M., & Clement, J. (2001). Evolving mental models of electric circuits. In Behrendt, H. et al. (Eds.), *Research in science education—past, present, and Future,* 235–240. Dordrecht: Kluwer.

van Zee, E., & Minstrell, J. (1997a). Using questioning to guide student thinking. *The Journal of the Learning Sciences, 6*(2), 227–269.

van Zee, E. H., & Minstrell, J. (1997b). Reflective discourse: Developing shared understanding in a physics classroom. *International Journal of Science Education, 19*(2), 209–228.

Chapter 2
An Instructional Model Derived from Model Construction and Criticism Theory

Mary Anne Rea-Ramirez
Western Governors University

John Clement
University of Massachusetts

Maria C. Núñez-Oviedo
Universidad de Conception

2.1 Introduction

In this chapter we present an overview of theories that led up to the current theory of model based teaching and learning, along with an introduction to model-based co-construction. We particularly recognize the role of conceptual change theories and social learning theories in the construction of understanding in science. In fact, these theories have played important roles in research and curriculum development. However, while each theory has contributed to progress in the field, each is limited in its ability to describe the cognitive processes that occur at many levels as a student and teacher co-construct understanding. In order to situate the research presented in this book, a brief description of each of these prior theories is included. This is not meant to be a full review but a brief introduction to provide a base for developing theories of model based learning and co-construction of knowledge. Although the background we bring to this work is rooted in the cognitive analysis of conceptual change processes, throughout this book we examine instructional dialogue in which social as well as cognitive methods are used to encourage students to build scientific models.

2.2 Classical Conceptual Change Theory

Classical Conceptual Change theory arose from the work of both Piaget's theories of children's thinking and Kuhn's history of science work, as well as work in science education on student's preconceptions, as shown in Fig. 2.1. The Piagetian theory of cognitive development suggested that disequilibrium or dissatisfaction within the individual must be created with the initial conception to produce learning. Posner, Strike, Hewson and Gertzog, (1982) proposed a model of conceptual change in students by analogy to Kuhn's description of the disequilibrium produced by anomalies in scientific revolutions (Kuhn, 1970). In addition, they related status to the intelligibility, plausibility, and fruitfulness of a new conception. In order to meet these four criteria, they suggest that the student must be dissatisfied with their present conception, must comprehend a proposed alternative conception, must believe that this alternative conception is plausible, and, finally, must see how using this alternative conception could help to solve problems or answer questions. Hewson and Hewson (1992) refer to these four conditions as determining the status of a conception for the learner and suggest that a change in the status will allow for a change in the concept. That is, when two competing conceptions both exist in the mind of an individual student, the relative status of each idea will determine which idea the student chooses to adopt.

Thagard (1992) also draws on the history and philosophy of science and cognitive psychology to propose a theory of conceptual change. Thagard speaks of degrees of conceptual change, stating that while some conceptual change simply involves addition, deletion, or revision of elements of knowledge, other types of conceptual change are more complex. He stresses that conceptual change is not merely a matter of revision of one's beliefs, since concepts are not simply collections of beliefs, but are "mental structures that are richly organized by means of relations such as kind and part (Thagard, 1992)". Larger changes would include major scientific revolutions in the natural sciences such as those involving Copernicus, Newton, Darwin, Einstein, quantum theory, and plate tectonics. Each of these, Thagard believes, involved major changes in conceptual organization involving kind and/or part relations (Thagard, 1992).

Classical conceptual change theory also grew out of the early finding from science education research that students come to the learning situation with many preconceptions, sometimes referred to by researchers as alternative conceptions, misconceptions, or naive conceptions, and that some of these preconceptions can be resistant to change. It is from this premise that researchers have developed strategies to foster and encourage conceptual change.

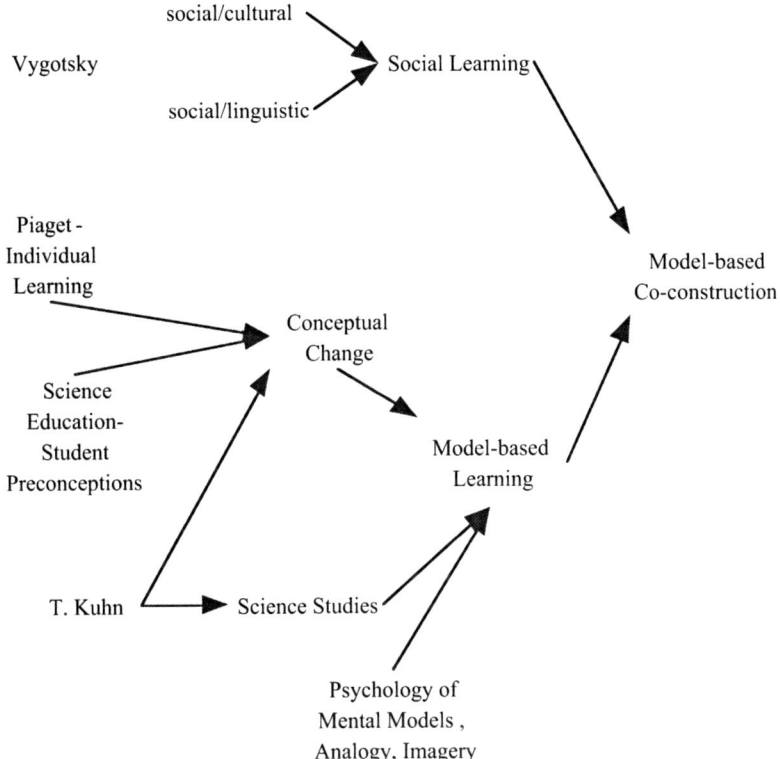

Fig. 2.1 Theories of learning in science

2.3 Social Learning Theory

While Piaget and many in the classical conceptual change field sought to describe the cognitive changes taking place within individual learners, Vygotsky (1978) suggested that knowledge is constructed by learners through social interaction, within the zone of proximal development which can be defined as "the distance between the actual developmental level as determined by independent problem solving and the level of potential development as determined through problem solving under adult guidance or in collaboration with more capable peers". That is, knowledge is negotiated as learners interact with each other and share ideas and experiences. This has been called social constructivism and stresses the social interaction and negotiation of new knowledge. This might occur at both the individual level and societal level. Among individuals in a group, new knowledge is accepted or rejected depending on the needs of the individuals to explain phenomena and communicate among themselves. Larger changes within societies depend on a filtering down of scientific understanding through education and possibly the mass media. Within social constructivist theory, there are multiple lenses, such as social-cultural and social linguistic.

The social-cultural lens sees knowledge as negotiated by and within communities. According to Cobern (1993) learners construct knowledge within the cultural context that gives meaning to that knowledge. Several concepts have been used in explaining the process, such as the zone of proximal development theory, expert scaffolding theory, Socratic dialogue theories, and reciprocal teaching.

Brown and Campione (1993) describe a type of discovery learning in which they conceive of instruction as "sequences of first small, then larger participatory groups, until the whole class is involved in what is called a community of learners." They state that it is teamwork that creates a "sub-culture of expertise" in which sequences of presentation, discussion, and deliberation play a major role in the construction and negotiation of understanding (Brown & Palincsar, 1989).

2.4 Limitations of Conceptual Change and Social Learning Theories

We distinguish between internal and external criticisms of classical conceptual change theory. External criticisms emphasize that cognitive methods used to promote conceptual change may be insufficient, failing to take into consideration motivational factors, the role of social learning, and the situational context of learning. Internal criticisms view classical conceptual change theory as flawed in that it emphasized too heavily big changes that occurred quickly and led to replacement rather than modification. Stavy (1991) and Dreyfus, Jungwirth and Eliovitch (1990) were concerned that sharp dissonance may discourage certain students. We would argue that most of the identified flaws internal to the classical theory of conceptual change theory can be repaired by appropriate modifications to the theory, and that external criticisms can be addressed by adding other types of teaching strategies. Clement (in press) believes a more serious internal problem was that the theory was not only flawed, but seriously underdeveloped internally; that is, it merely provided an outline for a theory of learning with some *conditions* for or *effects* of learning rather than a set of learning *mechanisms*.

On the other hand, sociocultural theories are very broad and often have little empirical support (Anderson, et al., 2001, p. 2). Some researchers have complained that these theories lack of specificity. "What gets internalized or appropriated and under what circumstances? How does the process work? One searches in vain for precise, data-based answers to these questions (Anderson, et al., 2001, p. 2)." In addition, one wonders whether social strategies alone can have an impact on persistent misconceptions. As an example, Leander and Brown (1999) documented classroom discussions in which impressive social skills for discussion were demonstrated, but very little conceptual change took place.

An obvious and important strategy is to merge the cognitive and social theories of learning for conceptual change (Hatano, 1993). Even though these theories are extremely valuable, they are still quite general and do not provide a sufficient understanding of underlying mechanisms to give much guidance for curriculum development. Hatano suggested that researchers have encountered two problems while attempting to merge sociocultural theory with the constructivist perspective. The first problem is the lack of more detailed descriptions about the interactions between the material provided by science teachers and an active mind. The second problem is the need to find better descriptions for explaining the way students' ideas change throughout the instruction. Both problems can be helped by mental modeling theory.

2.5 Mental Modeling Theory

Mental modeling theory can be viewed as a response to gaps in prior theories of reasoning and representation. First, the theory of mental models is opposed to a view of human thinking that is over reliant on rules of deductive reasoning. It emerged from a research tradition called "informal reasoning" (Voss, Perkins, & Segal, 1991). The field of informal reasoning examines alternatives to formal logic to describe thinking. In mental modeling theory, people build mental models that can be considered "structural analogs of real-world or imagined situations..." (Nersessian, 1992, p. 9).

The origins of mental modeling theories as an alternative view of reasoning can be traced to criticisms of positivist views of science made by historico-critical philosophers, such as Kuhn and Lakatos. These historians shifted their object of study from the products of science to the processes that create those products, and from a predominate focus on theory evaluation to include a focus on theory generation. Subsequent work in science studies has made further progress by examining data from original sources, such as scientific diaries and early manuscript drafts, that were instrumental in enormous contributions to the current understanding of physics (Nersessian, 1985, 1992, 1995; Tweney, 1985), genetics (Darden, 1991), and evolutionary theory (Gould, 1980; Gruber, 1974; Schweber, 1977). However, Tweney (2001) argued that even though Kuhn's ideas impacted the way sociologists and historians examined science, they had little impact on psychology.

It was not until recently that Kuhn's ideas aided in the emergence of a perspective that uses the "cognitive-historical" approach (Nersessian, 1987) to describe concept formation and conceptual change processes in science. Instead of studying the end point or products of science, Nersessian studied the extended process of conceptual change, including how scientific models are constructed and reconstructed. In order to explain the intermediate processes involved in the conceptual change, "this approach combines analyses of specific cases of conceptual change with analytical tools and theories of the cognitive sciences" (Nersessian, 1990, p. 34). The cognitive part of the method uses the psychological concept of a mental model to describe the core of scientific theories as a flexible and dynamic entity. The historical part of the approach is given by the examination of historical data, such as journals and papers of the physicist Maxwell (1831–1879), to provide case studies.

2.5.1 Explanatory Models as an Important Type of Scientific Model

Classic examples of models are images of atoms, molecules, the human circulatory system, black holes, swarms of particles in a gas, and the idea that the moon causes the tides. Nersessian gives a broad definition: "A mental model is a conceptual system representing the physical system that is being reasoned about. It is an abstraction – idealized and schematic in nature – that represents a physical situation by having surrogate objects and properties, relations, behaviors, or functions of these that are in correspondence with it." Gilbert and Boulter (2000) describe the human models as focusing us on certain features in a system. Models are useful when they capture important interrelationships in a system, as opposed to being a collection of isolated facts.

Historians of science such as Hesse (1967), and Harre (1972) have developed important distinctions between explanatory models, empirical law hypotheses, and formal principles, as shown in Table 2.1, adapted from Clement, Brown and Zietsman (1989). A particular subtype of model that is the main topic of focus in this book is the *explanatory model*. These hypothesized, theoretical, qualitative models, such as molecules, waves, and fields, are a separate kind of hypothesis from an observation pattern in the form of a graphical summary of the data collected in an experiment. An explanatory model is not simply a condensed summary of empirical observations but rather an invention that contributes new theoretical terms and images which are part of the scientist's view of the world, and which are neither "given" in nor implied by the data. Campbell's oft-cited example is that merely being able to make predictions from the empirical gas law stating that PV is proportional to RT, is not equivalent to understanding *why* they behave as they do in terms of an explanatory model of molecules in motion. The explanatory model provides a description of a hidden, non-observable process that explains how the gas works and answers "why" questions about where observable changes in temperature and pressure come from. (Summaries of these views are given in Harre, 1967; Hesse, 1967). Such an explanatory model can be thought of as an intermediate representation that stands between a data pattern and a formal (e.g. thermodynamic) model of the energy in a system of colliding gas particles. We can think of much of the core meaning of a student's theory of gasses as residing in such an explanatory model.

Table 2.1 Four types of knowledge used in science

	Types of knowledge	Example: study of gases
Theories	4. Formal Theoretical Principles	Principles of Thermodynamics
	3. Explanatory Models	Colliding elastic particle model
Observations	2. Qualitative or Mathematical Descriptions of Patterns in Observations, including Empirical Laws	Pv = kt (refers to observations of measuring apparatus)
	1. Primary-Level Data: Observations	Measurement of a single pressure change in a heated gas

2.5.2 Model Based Learning

A central question in our field, then, is: How do scientists and others construct scientific models? Historically, Nersessian (1995) points to Philip Johnson-Laird's theory of mental models as foundational. Johnson-Laird argued that people build mental models by using "the semantics of connectives rather than on rules of inference" (Johnson-Laird, 1986, p.34). He proposed that when an individual solves a logic problem, instead of using deductive reasoning, they may generate a model and check the accuracy through a process of investigating and eliminating alternative models that could rival the initial one (Johnson-Laird, 1983).

Other cognitive researchers have derived their descriptions of concept formation and conceptual change in science from examining the reasoning processes conducted by living experts while solving explanation problems (Clement, 1989) or while working in their lab (Dunbar, 1995; Nersessian, Kurz-Milcke, and Davies, 2005).

Another influence comes from work in cognitive science on the psychology of modeling processes such as analogy, mental model formation, and imagery use. Collins and Gentner (1987) proposed a theory of how people generate mental models by using analogies. These authors argued that individuals, when reasoning about simple unfamiliar domains, use analogical mapping "to create new mental models that they can then run to generate predictions about what should happen in various situations in the real world" (Collins & Gentner, 1987, p. 243). The known domain is called the "base" and the unknown domain is called the "target." When the unfamiliar situation is complex, "people partition the target system into a set of components models, each mapped analogically from a different base system" (Collins & Gentner, 1987, p. 248). In other words, they hypothesized that the subject can use more than one analogy to generate a complete mental model of an unknown situation. Other properties of mental models are discussed in Gentner and Stevens (1983) book that in some sense launched the field of the psychology of Mental Models. Clement (1988, 1989) documented the use of analogies by experts in constructing explanatory models while solving explanation problems.

More recently, some researchers have proposed that modelers can "run" dynamic models as mental simulations, allowing them to predict new findings for cases they have not seen before, almost as if they were running a "mental movie" from the model (Schwartz & Black, 1999; Clement, 2003, 2004; Trickett and Trafton, 2002; Craig, Nersessian and Catambone, 2002). They have indicated that certain geometric and analogical reasoning processes are easiest to explain by positing the use of imagistic mental simulation. If this were correct, it would seem to be a potentially powerful resource for science students. Clement (2003) has hypothesized that imagery can be transferred from a source analog to a target model.

2.6 Model Based Instructional Theory

Indeed it can be said that the early work on using analogies to foster conceptual change in instruction marks the beginning of model-based theories of instruction (Clement, 1993; Glynn, 1991; Harrison and Treagust, 1996). In the last decade there has been an increasing recognition of the role of models and modeling in describing instructional strategies (Chiu, Chou, & Liu, 2002; Gilbert & Boulter, 1998; J. D. Gobert & Buckley, 2000; Schnotz & Preub, 1999). Mental models have led to the identification of important factors in learning and have supported curriculum reform efforts with some remarkable successes (Chi, Slotta, & Leeuw, 1994; Clement, 1993c; Clement, Brown, & Zietsman, 1989; Hennessey, 1999; Johnson & Stewart, 1990; Rea-Ramirez, 1998; P. H. Scott, Asoko, & Driver, 1992; Steinberg & Wainwright, 1993; White & Frederiksen, 1998; Raghavan and Glaser, 1995). Although several studies acknowledge that students build mental models when attempting to understand a system (Driver, 1995; Duit & Treagust, 1998; Hogan, Nastasi, & Pressley, 2000; D. Kuhn, Black, Keselman, & Kaplan, 2000) or analyze how this happens by using cycles of model construction and criticism (Buckley, 2000; Clement & Steinberg, 2002; J. Gobert, 2000; Rea-Ramirez, 1999), these studies do not interpret teacher-student classroom dialogue in terms of the learner's construction and revision of mental models. As a consequence, we still lack a clear set of mechanisms to explain how students build mental models about a target concept, how to deal with multiple alternative conceptions in large classroom settings, and how to connect teacher-student interactions with mental models. In short, we still lack a detailed theory of model-based learning and teaching (MBTL) (Clement, 2000b; Gobert & Buckley, 2000).

What is needed, then, is work in the field that integrates the key ideas of dissonance and conceptual change, with new model based perspectives on the kinds of concrete representations involved in knowledge construction and the role of abduction, analogy, and other strategies in forming such models. This cognitive theory then must be set in a realistic social context of the classroom where social interactions are critical in determining whether such productive cognitive processes take place. Thus, we hope to make a contribution to integration in the field by placing cognitive theory in a social context. Driver and her colleagues have called for this kind of integration:

> Learning science involves young people entering into a different way of thinking about and explaining the natural world; becoming socialized to a greater or lesser extent into the practices of the scientific community with its particular purposes, ways of seeing, and ways of supporting its knowledge claims. Before this can happen, however, individuals must engage in a process of personal construction and meaning making. Characterized in this way, learning science involves both personal and social processes (Driver, Asoko, Leach, Mortimer, & Scott, 1994, p. 8).

Due to limitations of time and space, in this book we concentrate on the social context of student-teacher interactions in whole class discussions–student-teacher co-construction. For now we must leave the analysis of small group student-student interactions to another time and place, although each of the curricula we discuss depend heavily on both large and small group interactions.

2.6.1 Introduction to Model-based-co-construction

Building on the theories reviewed above, in this book we will focus on a strategy called "Model-Based Co-construction" that attempts to integrate cognitive and social elements. Some social constructivist researchers recently have also used the term co-construction for describing teacher-student or student-student interactions and guided instruction processes (Billett, 1996; Bulgren, Deshler, Schumaker, & Lenz, 2000; Driver, Asoko, Leach, Mortimer and Scott, 1994; Hogan, 1999b). However most do not provide a definition for the term and none interpret teacher-student interactions in terms of the learner's construction and revision of mental models.

In this section we will review recent papers by our own group that have led up to the results reported in this book. We use the term co-construction to describe the process by which the teacher and the students both contribute ideas to build and evaluate a model (Rea-Ramirez, 1998; Nunez-Oviedo, 2001; Nunez-Oviedo, Rea-Ramirez, Clement, & Else 2002; Nunez-Oviedo & Clement 2002). Chapter 1 introduced an example of an episode of co-construction in the area of pulmonary respiration. The approach may be considered a middle road between purely teacher generated and purely student generated ideas in the classroom. In the view to be presented, the process of constructing a model is not the result of a sudden large change in the student's model but the result of many small episodes, often initiated by what we call "dissatisfaction" or dissonance. Clement and Rea-Ramirez (1998) reviewed previous theories and synthesized an extended theory of dissonance and dissatisfaction. This view has origins in Festinger's theory of cognitive dissonance (Harmon-Jones & Mills, 1999) as well as Piaget's equilibration theory. They proposed that dissonance could be used to trigger smaller, milder revisions in a model in contrast to some who have assumed that the use of dissonance can only be used to reject a prior model leading to a large conceptual change.

Dissonance was defined as "an internal sense of disparity between an existing conception and some other entity. This can occur at mild, as well as strong levels, as opposed to the concept of 'conflict,' which suggests only a strong disparity" (Clement & Rea-Ramirez, 1998, p. 2). They proposed that an existing conception might be compared with external sources of dissonance such as discrepant events, analogies, and counterexamples and with internal sources such as incoherence between two conceptions. They suggested that a source of dissonance must be recognized and internalized by the students before the actual internal dissonance may take place (Chinn & Brewer, 1998).

Along with Clement and Steinberg (2002), they recognized that another reason students sometimes become dissatisfied with their current model is because they sense a gap in what it can explain – it provides no explanation for some events. These authors proposed the term "dissatisfaction" to include both possibilities – the sensed presence of a "gap" in one's ability to explain, or the sensed presence of a conflict between two ideas (dissonance). In the remainder of this chapter, the term dissatisfaction will be used to describe the sense of mild or strong discomfort that the student may feel in either of these situations.

2.6.2 Model Evolution

It was hypothesized that small model revisions may be motivated by using one or more episodes of dissatisfaction. A number of dissatisfaction episodes and revisions may be needed depending on the distance between the initial model and the target model, as shown in Fig. 2.2. This *model evolution* process originates increasingly sophisticated models until reaching the target concept. Clement (2000b) focused on situations where instructional efforts were directed at helping a student move from model M_n to model M_{n+1} (Clement, 2000b, p. 1043). The resulting sequence of intermediate mental models is also called a "learning pathway" (Rea-Ramirez, 1998; Scott, 1992).

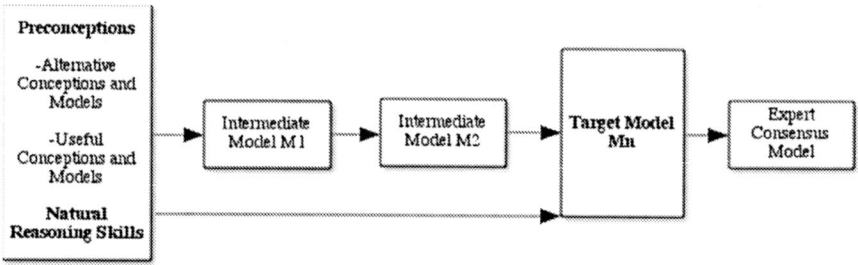

Fig. 2.2 Learning pathway

While the learning pathway is the outcome of a model evolution process, it does not describe the process that generates model evolution. Contemporary science historians and cognitive researchers developed the idea that scientists generate their hypotheses by process that has been called "constructive modeling" (Nersessian, 1995) or "generation, evaluation, and modification cycles" or GEM cycles (Clement, 1989, 1993a, 1993b).

Constructive modeling "is a process of abstracting and integrating constraints into successive models of the target problem" (Nersessian, 1995, p. 222). Figure 2.3 depicts a GEM cycle derived from expert protocols that illustrates a cyclical process of hypothesis generation, rational and empirical testing, and modification or rejection (Clement, 1989). In other words, GEM cycles are processes in which an initial model is criticized and then revised, originating a series of models with an increasing complexity and sophistication.

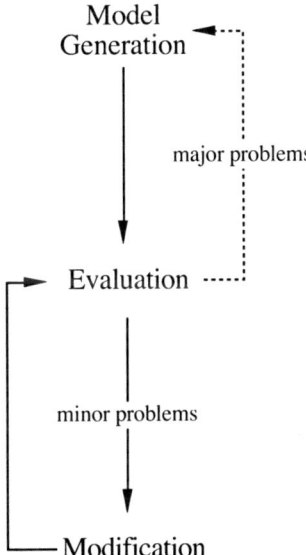

Fig. 2.3 GEM cycle (adapted from Clement, 1989)

Since this cycle can use non-formal rather than formal reasoning processes, it should be possible for students to engage in it. Clement (1989) and Nersessian (1992, 1995) both predicted that the thinking practices in which scientists and experts engage, such as constructive modeling and cycles of model construction and criticism should be directly relevant to student learning but they did not provide guidelines to teachers and curriculum developers on how to guide instruction.

2.7 Conclusion

This book is then largely an attempt to bridge the gap from these theories of modeling processes to practical implications for education. There are two major pieces that form the bridge: (1) the specification of viable intermediate model sequences (learning pathways) in specific instructional areas; and (2) the providing of an expanded pedagogy and set of teaching strategies that are appropriate for the model construction cycles shown above. The background work behind these pieces is as follows.

2.7.1 Determining Target Models and an Effective Learning Pathway

An important difficulty with interpreting the current science standards in the USA is that they only specify a target concept or model. The specification of the learning pathway – of intermediate models, common misconceptions, and possible positive preconceptions to build on – is missing. Examples of these are needed in order to guide future work. Clement, Brown, and Zietsman (1987) and Niedderer and Goldberg (1996) developed early examples of learning pathways through tutoring studies in mechanics and electricity. Rea-Ramirez (1998) conducted an extensive study of human respiration ideas through in-depth interviews and tutoring studies with individual students who had had varying amounts of instruction. She uncovered a detailed learning pathway observed in students who were able successfully to construct an integrated and dynamic understanding of respiration at both the cellular and organismic level. Portions of this work are described in Chapter 3, and Steinberg describes a pathway for introductory electricity in Chapter 5. These examples illustrate the form that a detailed learning pathway can take.

2.7.2 Model Based Pedagogy

We need to develop an expanded pedagogy and set of teaching strategies for fostering model construction in science classrooms. Early work on this problem by others was reviewed earlier. Within our own group, Clement and Steinberg (2002) and Rea-Ramirez (1998) studied teaching strategies in both tutoring and small group contexts. In the Rea-Ramirez respiration study, a large number of "mini-cycles" of model construction were observed to occur throughout the intervention supported by multiple strategies, some that were both unexpected and unplanned prior to the intervention. During mini-cycles students returned to their prior model much more frequently than expected and made smaller changes in those models than expected (Rea-Ramirez, 1998). From tutoring studies, new models of learning were used to develop strategies for teaching about respiration that were tested with small groups and eventually whole class instruction (Rea-Ramirez, 1998; Rea-Ramirez, 1999; Rea-Ramirez & Gibson, 2001; Rea-Ramirez, Nunez-Oviedo, Clement, Else, 2002).

Nunez-Oviedo (2003) investigated strategies used in full classrooms while building the same respiration ideas. Building on Ramirez' minicycles in individuals, she observed recurring micro cycles in large group discussions. A Micro Cycle is a small teacher-student interaction pattern that consists of two or more intertwined "cycles of generation, evaluation, and modification", or GEM cycles, working together. Ideally, the teacher acts mostly to encourage the cycle, while the students actually generate ideas in the cycle. Micro Cycles can modify or disconfirm model elements or entire models. Several larger pedagogical patterns and a variety of instructional strategies observed are discussed by Nunez in Chapter 7 and 10, and examples from different subject areas are discussed in other chapters. We will describe observed teacher-student interaction patterns in a spirit of inquiry similar to the methodology used in Resnick, Salmon, Seits, Wathen, and Holowchak (1993). By analyzing transcripts in detail, they were able to build models of how students reacted to each other's statements during discussion and illustrate ways in which the conversation was coherent and supported learning in terms of a process of shared reasoning. While their work focused on student argumentation patterns, our work also includes teacher-student interactions. The present work develops the analysis of specific teaching strategies at different time scale levels and develops diagrammatic notations for them via methods described in Clement (2000a). These processes can be described in terms of the construction of mental models via dialectic interactions that involve partial models, blind alleys, criticisms, and revisions, and incremental changes toward conceptual change. In other words, we examine co-construction from the perspective of reasoning and learning processes.

Although the background we bring to this work is rooted in the cognitive analysis of conceptual change processes in students and scientists, throughout this book we examine instructional dialogue in which socio-cognitive methods are used to encourage students to build scientific models. This can be seen as a contribution to the efforts to integrate social and cognitive perspectives for explaining science instruction. In this way we hope is to contribute elements to a theory of model-based co-construction by examining common patterns observed within middle school, high school and college science classes in several science disciplines.

References

Anderson, R. C., Nguyen-Jahiel, K., McNurlen, V., Archodidoi, A., Kim, S.-y, Reznitskaya, A., et al. (2001). The snowball phenomenon: Spread of ways of talking and ways of thinking across groups of children. *Cognition and Instruction, 19*, 1–46.

Billett, S. (1996). Situated learning: Bridging sociocultural and cognitive theorizing. *Learning and Instruction, 6(3)*, 263–280.

Brown, A., & Campione, J. (1993). Guided discovery in a community of learners. In K. McGilly (Ed.), *Classroom lessons: Integrating cognitive theory and classroom practice*. Cambridge: MIT Press.

Brown, A. L., & Palincsar, A. S. (1989). Guided, cooperative learning and individual knowledge acquisition. In L. B. Resnick (Ed.), *Knowing, learning, and instruction. Essays in honor of Robert Glaser* (pp. 393–451). Hillsdale, New Jersey: Lawrence Erlbaum Associates Publishers.

Buckley, B. C. (2000). Interactive multimedia and model-based learning in biology. *International Journal of Science Education, 22(9)*, 895–935.

Bulgren, J. A., Deshler, D. D., Schumaker, J. B., & Lenz, B. K. (2000). The use and effectiveness of analogical instruction in diverse secondary content classrooms. *Journal of Educational Psychology, 92(3)*, 426–441.

Chi, M. T. H., Slotta, J. D., & Leeuw, N. d. (1994). From things to processes: A theory of conceptual change from learning science concepts. *Learning and Instruction, 4*, 27–43.

Chinn, C. A., & Brewer, W. F. (1998). An empirical test of a taxonomy of responses to anomalous data in science. *Journal of Research in Science Teaching, 35*, 623654.

Chiu, M.-H., Chou, C.-C., & Liu, C.-J. (2002). Dynamic processes of conceptual change: Analysis of constructing mental models of chemical equilibrium. *Journal of Research in Science Teaching, 39(8)*, 688–712.

Clement, J. (1988). Observed methods for generating analogies in scientific problem solving. *Cognitive Science, 12*, 563–586.

Clement, J. (1989). Learning via model construction and criticism. In G. Glover, R. Ronning, & C. Reynolds (Eds.), *Handbook of creativity: Assessment, theory and research* (pp. 341–381). New York: Plenum.

Clement, J. (1993a). *Model construction and criticism cycles in expert reasoning.* Paper presented at the Fifteenth Annual Meeting of the Cognitive Science Society, Hillsdale, NJ.

Clement, J. (1993b). *Scientific learning in experts: Explanatory model construction vs. induction from observations.* Paper presented at the AERA Conference.

Clement, J. (1993c). Using bridging analogies and anchoring intuitions to deal with students' preconceptions in physics. *Journal of Research in Science Teaching, 30(10)*, 1241–1257.

Clement, J. (2000a). Analysis of clinical interviews: Foundations and model viability. In A. E. Kelly, & R. Lesh (Eds.), *Handbook of research methods in mathematics and science education* (pp. 547–589). Mahwah, NJ: Lawrence Erlbaum Associates.

Clement, J. (2000b). Model based learning as a key research area for science education. *International Journal of Science Education, 22(9)*, 1041–1053.

Clement, J. (2003). Imagistic simulation in scientific model construction. In R. Alterman and D. Kirsh, Editors, *Proceedings of the Twenty-Fifth Annual Conference of the Cognitive Science Society, 25,* 258–263. Mahwah, NJ: Erlbaum.

Clement, J. (2004). Imagistic processes in analogical reasoning: Conserving transformations and dual simulations. In K Forbus, D. Gentner & T. Regier, (Eds.), *Proceedings of the Twenty-Sixth Annual Conference of the Cognitive Science Society, 26,* 233–238. Mahwah, NJ: Erlbaum.

Clement, J. (in press). The role of explanatory models in teaching for conceptual change. In S. Vosniadou, (Eds.), *Handbook of research on conceptual change.*

Clement, J., Brown, D. E., & Zietsman, A. (1989). Not all preconceptions are misconceptions: Finding 'anchoring conceptions' for grounding instruction on students' intuitions. *International Journal of Science Education, 11*(Special Issue), 554–565.

Clement, J., & Rea-Ramirez, M. A. (1998). The role of dissonance in conceptual change. *Proceedings of the National Association for Research in Science Teaching Conference.*

Clement, J., & Steinberg, M. (2002). Step-wise evolution of mental models of electric circuits: A "learning-aloud" case study. *Journal of the Learning Sciences, 11(4),* 389–452.

Cobern, W. W. (1993). Contextual constructivism: The impact of culture on the learning and teaching of science. In K. Tobin (Ed.), *The practice of constructivism in science education.* Washington, DC: American Association for the Advancement of Science, pp. 51–69.

Collins, A., & Gentner, D. (1987). How people construct mental models. In D. Holland & N. Quinn (Eds.), *Cultural models in thought and language* (pp. 243–265). Cambridge, UK: Cambridge University Press.

Craig, D. L., Nersessian, N., & Catambone, R. (2002). Perceptual Simulation in Analogical Problem Solving. In L. Magnani, & N. Nersessian (Eds.), Model-Based Reasoning: Science, Technology, Values (pp. 167–189). New York: Kluwer Academic Publishers.

Dreyfus, A., Jungwirth, E. and Eliovitch, R. (1990) Applying the 'Cognitive Conflict' strategy for conceptual change - some implications, difficulties and problems. *Science Education,* 74 (5), 555–569

Darden, L. (1991). *Theory change in science. Strategies from Mendelian genetics.* Oxford, UK: Oxford University Press.

Driver, R. (1995). Constructivist approaches to science teaching. In L. P. Steffe & J. Gale (Eds.), *Constructivism in education* (pp. 385–400). Hillsdale, NJ: Lawrence Erlbaum Associates.

Driver, R., Asoko, H., Leach, J., Mortimer, E., & Scott, P. (1994). Constructing scientific knowledge in the classroom. *Educational Researcher, 23(7),* 5–12.

Duit, R., & Treagust, D. F. (1998). Learning in science - from behaviorism towards social constructivism and beyond. In B. J. Fraser & K. G. Tobin (Eds.), *International handbook of science education* (pp. 3–25). Great Britain: Kluwer Academic Publishers.

Dunbar, K. (1995). How scientist really reason: scientific reasoning in real-world laboratories. In R. J. Sternberg & J. Davidson (Eds.), *The nature of insight* (pp. 256–396). Cambridge, MA: MIT Press.

Gentner, D., & Stevens, A. L. (Eds.) (1983). *Mental models.* Hillsdale, NJ: Lawrence Erlbaum Associates.

Gilbert, J. K., & Boulter, C. (1998). Learning science through models and modeling. In B. J. Fraser & K. G. Tobin (Eds.), *International handbook of science education* (pp. 53–66). UK: Kluwer Academic Publishers.

Gilbert, J. K. & Boulter, C. J. (2000). Developing models in science education. Dordrecht, The Netherlands: Kluwer Academic Publishers.

Glynn, S. M. (1991). Explaining science concepts: A teaching-with-analogies model. In Glynn, S. M., Yeany, R. H., & Britton, B. K. (Eds.) *The Psychology of Learning Science* (pp. 219–240). Hillsdale, NJ: Erlbaum.

Gobert, J. (2000). A typology of causal models for plate tectonics: Inferential power and barriers to understanding. *International Journal of Science Education, 22*(9), 937–977.

Gobert, J. D., & Buckley, B. C. (2000). Introduction to model-based teaching and learning in science education. *International Journal of Science Education, 22*(9), 891–894.

Gould, S. J. (1980). Darwin's middle road. In *The panda's thumb. More reflections in natural history.* New York: W. W. Norton & Company.

Gruber, H. (1974). *Darwin on man.* New York: E. P. Dutton.

Harmon-Jones, E. & Mills, J. (1999). Cognitive dissonance: Progress on a pivotal theory in social psychology. Appendix B.

Harré, R. (1972). The philosophies of science: An introductory survey. NY: Oxford University Press.

Harrison, A. G., & Treagust, D. F. (1996). Secondary students' mental models of atoms and molecules: implications for teaching science. Science Education, 80, 509–534.

Hatano, G. (1993). Commentary: Time to merge Vygotskian and constructivist conceptions of knowledge acquisition. In E. A. Forman, N. Minick & C. A. Stone (Eds.), *Contexts for learning: Sociocultural dynamics in children's development.* New York: Oxford University Press.

Hennessey, G. M. (1999). *Probing the dimensions of metacognition: Implications for conceptual change teaching-learning.* Paper presented at the NARST Conference, Boston, MA.

Hesse, M. (1966). Models and Analogy in Science. South Bend, IN: Notre Dame University Press.

Hewson, P., & M. Hewson. (1992). The status of students' conceptions. In R. Duit, F. Goldberg, & H. Niedderer. (Eds.), *Research in Physics Learning: Theoretical Issues and Empirical Studies*, Kiel, Germany: Institute for Science Education.

Hogan, K. (1999b). Relating students' personal frameworks for science learning to their cognition in collaborative contexts. *Science Education, 83*, 1–32.

Hogan, K., Nastasi, B. K., & Pressley, M. (2000). Discourse patterns and collaborative scientific reasoning in peer and teacher-guided discussions. *Cognition and Instruction, 17(4)*, 379–432.

Johnson, S. K., & Stewart, J. (1990). Using philosophy of science in curriculum development: An example from high school genetics. *International Journal of Science Education, 12(3)*, 297–307.

Johnson-Laird, P. N. (1983). *Mental models*. Cambridge, MA: Harvard University Press.

Johnson-Laird, P. N. (1986). Reasoning without logic. In T. Myers, K. Brown & B. McGonigle (Eds.), *Reasoning and discourse processes* (pp. 13–49). London, UK: Academic Press.

Kuhn T. S. (1970) The Structure of Scientific Revolutions. Chicago: Chicago University Press.

Kuhn, D., Black, J., Keselman, A., & Kaplan, D. (2000). The development of cognitive skills to support inquiry learning. *Cognition and Instruction, 18*(4), 495–523.

Leander, K. M., & Brown, D. E. (1999). "You understand but you don't believe it": Tracing the stabilities and instabilities of interaction in a physics classroom through a multidimensional framework. *Cognition and Instruction, 17(1)*, 93–135.

Niedderer, H., Goldberg, F. (1996). Learning Processes in Electric Circuits. Paper presented at NARST, St. Louis MO, USA. (http://didaktik.physik.uni-bremen.de/niedderer/pubs.htm)

Nersessian, N. J. (1985). Faraday's field concept. In D. Gooding & F. A. J. L. James (Eds.), *Faraday rediscovered. Essays on the life and work of Michael Faraday 1791–1867* (pp. 175–187). New York, NY: Stockton Press.

Nersessian, N. J. (1987). A cognitive-historical approach to meaning in scientific theories. In N. J. Nersessian (Ed.), *The process of science* (pp. 161–177). Dordrecht: Martinus Nijhoff Publishers.

Nersessian, N. J. (1990). Methods of conceptual change in science: Imagistic and analogical Reasoning. *Philosophica, 45*, 33–52.

Nersessian, N. J. (1992). How do scientists think? Capturing the dynamics of conceptual change in science. In R. N. Giere (Ed.), *Cognitive models of science* (Vol. 15, pp. 3–44). Minneapolis: University of Minnesota Press.

Nersessian, N. J. (1995). Should physicists preach what they practice? Constructive modeling in doing and learning physics. *Science & Education, 4*, 203–226.

Nersessian, N. J., Kurz-Milcke, E., & Davies, J. (2005). Ubiquitous computing in science and engineering research laboratories: A case study from biomedical engineering. In G. Kouzoulis et al., (Eds.), (pp. 167–198). Knowledge in the New Technologies Berlin: Peter Lang Publishers.

Nunez-Oviedo, M. C. (2001). *A teaching method derived from model construction and criticism theory.* Unpublished Comprehensive Exam Paper, University of Massachusetts, Amherst.

Nunez-Oviedo, M. C. (2003). Teacher-student co-construction processes in biology: Strategies for developing mental models in large group discussions. Unpublished Doctoral Dissertation, University of Massachusetts, Amherst, MA.

Nunez-Oviedo, M. C., & Clement, J. (2002). *An instructional method derived from model construction and criticism theory.* Paper presented at the NARST Conference, New Orleans, LA.

Nunez-Oviedo, M. C., Rea-Ramirez, M. A., Clement, J., & Else, M. J. (2002). Teacher-student co-construction in middle school life science. *Proceedings of the AETS Conference.*

Posner, G. J., Strike, K. A., Hewson, P. W., & Gertzog, W. A. (1982). Accommodation of a scientific conception: Toward a theory of conceptual change. *Science Education, 66(2),* 211–227.

Raghavan, K. & Glaser, R. (1995). Model based analysis and reasoning in science: the MARS curriculum. *Science Education,* 79, 37–61.

Rea-Ramirez, M. A. (1998). Model of conceptual understanding in human respiration and strategies for instruction. (Doctoral Dissertation) *DAI - 9909208, University of Massachusetts, Amherst.*

Rea-Ramirez, M. A. (March, 1999). *Developing complex mental models through explanatory need.* Paper presented at the NARST Conference, Boston, MA.

Rea-Ramirez & Gibson, (2001). Keeping the inquiry in curriculum designed to help students' conceptual understanding of cellular respiration. *Proceedings of the 2002 Annual International Conference of the Association for the Education of Teachers in Science.* Charlotte, N. C.

Rea-Ramirez, M. A., Nunez-Oviedo, M. C., Clement, J., Else, M. J. (2002). Energy in the human body: a middle school curriculum. Amherst, MA: University of Massachusetts.

Resnick, L. B., Salmon, M., Seitz, C. M., Wathen, S. H., & Holowchak, M. (1993). Reasoning in conversation. *Cognition and Instruction, 11(3 & 4),* 347–364.

Schnotz, W., & Preub, A. (1999). Task-dependent construction of mental models as a basis for conceptual change. In G. Rickheit, & C. Habel (Eds), *Mental models in discourse processing and reasoning.* Amsterdam: North-Holland.

Schwartz, D. L., & Black, T. (1999). Inferences through imagined actions: knowing by simulated doing. *Journal of Experimental Psychology: Learning, Memory, and Cognition,* 25, 116–136.

Schweber, S. (1977). The origin of the origin revisited. *Journal of the History of Biology, 10*(229–316).

Scott, P. H. (1992). Pathways in learning science: A case study of the development of one student's ideas relating to the structure of matter. In R. Duit, F. Goldberg & H. Niederer (Eds.), *Research in physics learning: Theoretical issues and empirical studies. Proceedings of an International Workshop held at the University of Bremen* (pp. 203–224). Kiel, Germany: IPN/Institute for Science Education.

Scott, P. H., Asoko, H. M., & Driver, R. H. (1992). Teaching for conceptual change: A review of strategies. In R. Duit, F. Goldberg & H. Niederer (Eds.), *Research in physics learning: Theoretical issues and empirical studies. Proceedings of an International Workshop held at the University of Bremen* (pp. 310–329). Kiel, Germany: IPN/Institute for Science Education.

Stavy, R. (1991) Using analogy to overcome misconceptions about conservation of matter. *Journal of Research in Science Teaching 28 (4),* 305–313

Steinberg, M., & Wainwright, C. L. (1993). Using models to teach electricity -the CASTLE project. *The Physics Teacher, 31(6),* 353–357.

Trickett, S. & Trafton, J. G. (2002) The instantiation and use of conceptual simulations in evaluating hypotheses: Movies-in-the-mind in scientific reasoning. In WayneGray and Christian Schunn, (Eds.), *Proceedings of the Twenty-Fourth Annual Conference of the Cognitive Science Society* 22, 878–883. Mahwah, NJ: Erlbaum.

Thagard, P. (1992). *Conceptual Revolutions*, Princeton, NJ, Princeton University Press.

Thagard, P. (1997) Coherent and Creative Conceptual Combinations. In: T. B. Ward, S. M. Smith, and J. Viad (eds), Creative thought: an investigation of conceptual structures and processes. Washington, DC. American Psychological Association.

Tweney, R. D. (1985). Faraday's discovery of induction: A cognitive approach. In D. Goodling & F. James (Eds.), *Faraday rediscovered: Essays on the life and work of Michael Faraday, 1791–1867* (pp. 189–209). New York: Stockton Press.

Tweney, R. D. (2001). Scientific thinking: A cognitive-historical approach. In K. Crowley, C. D. Schunn & T. Okada (Eds.), *Designing for science. Implications from everyday, classroom, and professional settings* (pp. 141–173). Mahwah, NJ: Lawrence Erlbaum Associates.

Voss, J. F., Perkins, D. N., & Segal, J. W. (Eds.). (1991). *Informal reasoning and education*. Hillsdale, NJ: Lawrence Erlbaum Associates.

Vygotsky, L. S. (1978). *Mind and society: The development of higher mental processes*. Cambridge, MA: Harvard University Press.

Vygotsky, L. (1986). *Thought and language*. Cambridge, MA: MIT Press.

White, B. Y., & Frederiksen, J. R. (1998). Inquiry, modeling, and metacognition: Making science accessible to all students. *Cognition and Instruction, 16(1)*, 3–118.

Zietsman, A. and Clement, J. (1987). The role of extreme case reasoning in instruction for conceptual change. *Journal of the Learning Sciences, 6(1)*, 61–89.

Section II

Introduction to Model Based Teaching Strategies

Chapter 3
Determining Target Models and Effective Learning Pathways for Developing Understanding of Biological Topics

Mary Anne Rea-Ramirez

Western Governors University

3.1 Introduction

Prior research has indicated that students of all ages show little in depth understanding of complex biological topics that form the building blocks for further study (Bishop, Roth, & Anderson, 1986; Sanders, 1993; Songer & Mintzes, 1994). This occurs even after instruction, with alternative conceptions persisting into adulthood. Whether this is due to faulty educational strategies leading to difficulty in understanding a complex system is an important question. In this chapter I argue that examining how much and what type of understanding is necessary to construct a worthwhile comprehension of complex topics is a critical first step in developing instructional strategies for teaching these difficult topics in biology. This first entails determining target models and a learning pathway that provide realistic chances for understanding (building on the concepts of target model and intermediate model developed in Chapter 1). This then allows the instructor to develop a set of strategies that aid the student in moving along this learning pathway, from Intermediate Models to Intermediate Models, gradually evaluating and modifying their initial conception (Rea-Ramirez, 1998; Niedderer and Goldberg, 1996). Camp, et al. (1994) used diagrams to convey similar pathways for physics teachers. In this chapter I attempt to develop a general framework for thinking about model-based curriculum goal structures. This framework is abstracted from our experiences in conducting diagnostic research and formative evaluation while developing a middle school life sciences curriculum, The Energy in the Human Body Curriculum. I will discuss how to identify effective target models and develop learning pathways for effective teaching. The Energy Curriculum involved the use of a learning pathway previously observed in students who were able successfully to construct an understanding of respiration at both the cellular and organismic level.

3.2 Target Concepts

Three basic concerns in children's learning of science have been identified (Chiu, & Leeuw, 1991). First, a significant number of children's ideas about science undergo no change in response to teaching or if they do change, they are not in the direction intended by the teacher. Second, these alternative conceptions held by children often persist even into adulthood and, third, alternative conceptions are not just random but can actually develop out of ideas that are both coherent and integrated even though they are incorrect from a scientific point of view. These concerns appear to be exacerbated in biology, where American students ranked last among 16 nations in achievement levels in biology and where an understanding of issues in biology is imperative for students to interact in the world around them in a constructive way.

Addressing the challenge of conceptual change in biology, particularly in the area of complex systems is a very large and challenging task. Research in science education has indicated that alternative conceptions in science are often very difficult to overcome (Bishop, Roth, & Anderson, 1986; Sanders, 1993; Songer & Mintzes, 1994). With this in mind, which concepts are determined to be important and how instruction is structured becomes central. However, when expectations of learning are developed from the expert's view, it may be that we are setting students up for failure. To address this problem, researchers have suggested many strategies such as analogies, discrepant events, hands-on learning activities. In this chapter, however, the discussion will concern itself more with how the original target concepts and learning pathways are determined, rather than specific teaching strategies used to move students along this pathway. It appears that this is an important first step that will ultimately aid the instructor in selecting the best and most effective strategies to use.

3.2.1 Why Identify Target Concepts

The problem of persistent alternative and naive conceptions exists in regards to human respiration as well as in many areas of science. Biology learning and teaching research suggests that students have difficulty developing deep conceptual understanding when a more traditional approach to teaching is taken, that is, one in which structures are learned in detail first, then function is introduced, and little connection between systems is made (Nagy, 1953; Gellert, 1962; Porter, 1974; Tait & Ascher, 1955). However, as alternative means of assisting students to construct understanding continued to be sought, they often entailed lessons that attempted to cause students to make giant leaps from preconception to scientific conception without taking into account the difficulty and distance of this leap, as well as how deeply imbedded and resistant the preconceptions were. In addition, the intended outcome sought by the instructor is often not clearly articulated and may not be consistent with student developmental levels or ability.

For this reason, there appears to be a need for establishing realistic goals for instruction that take into account the students' preconceptions, misconceptions, and ability level. This involves the identification of target concepts that are developmentally and academically appropriate, that are aligned with National Standards, and are scientifically sound. This combination allows instructors to provide guidance to students as they move along a learning path to achieve a realistic goal. Rather than have one large goal for a lesson or series of lessons, the instructor can set targets that act as steps along a learning pathway. From these realistic target concepts the instructor can then develop a series of appropriate strategies that will allow the students to gradually criticize and modify their models, moving toward the target concept.

3.2.2 Levels of Target Concepts

To be scientifically sound does not necessarily mean that every target concept is at the expert scientific level. For instance, it would be unrealistic to expect seventh grade students to understand how the mitochondria work from an expert's point of view. They would not have the necessary background in biology and chemistry. However, students could develop a basic understanding of the mitochondria including the products needed to transform energy. The final target concept is determined by the academic and developmental level of the students. In this chapter I will use the term "concept" interchangeably with the term "model".

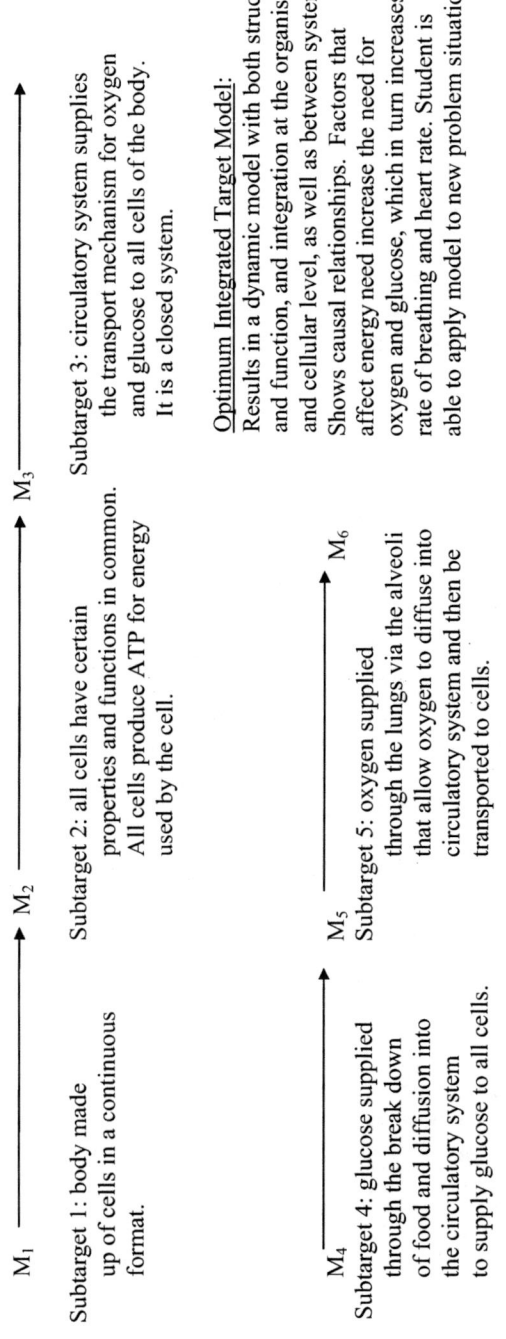

Fig. 3.1 Linear representation of Target Models and Integrated Multi-Target Structure for Entire Curriculum showing breakdown into stages or Target Models needed to help students move from their initial model to a more scientifically complete model

In addition, along a learning pathway leading to a complex biological concept, there may be multiple smaller target concepts. That is, in the case of students developing a complex conceptual model of respiration, the instructor may identify multiple smaller sub-target concepts that act as stepping stones for the student as they construct the larger more complex final model. For example, in the Energy in the Human Body Curriculum, there is the main target concept, referred to in Fig. 3.1 as the *Integrated Multi-Target Structure for Entire Curriculum* (IMTS). This represents the final target concept (IMTS) for the entire unit. However, it would be unrealistic to assume that a student could move from an initial model (M1) or preconception about how energy works in the body, including extensive information about systems and their interactions, in response to one activity, one instructor presented model or a single lecture.

For this reason, multiple target concepts are used to break the process of constructed understanding into more reachable goals. The target models are in themselves quite large in this particular curriculum as is the case in many life science areas. Figure 3.1 shows the five target concepts that make up the pathway to the final target model. As can be seen from the description of the final target, while integrated and dynamic, this is not an expert model but one that was determined appropriate for a middle school student. It is believed that this level of understanding would provide a firm foundation upon which students could further build during high school and college biology.

While the Energy Curriculum provides instructors with a series of lessons to help students criticize and revise or construct the target concepts, this process is often not available in current published curriculums adopted by many school districts. Therefore, while the final target, activities, and factual information may be given, it will fall to the instructor to develop the appropriate target models prior to deciding on the best strategies to use. As in the case with the Energy Curriculum, even the target models may be large, complex concepts. The target must also be broken into multiple steps to allow the students time to criticize and revise their models. The instructor may also need to develop multiple intermediate models concepts along the way to developing the target. We can represent this nested set of target concepts as a tree, part of which is shown in Fig. 3.2.

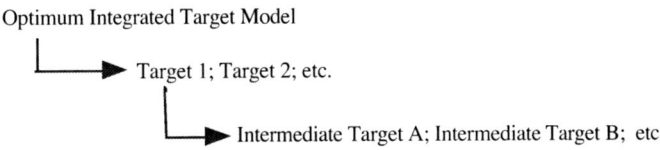

Fig. 3.2 Level of target concepts

Small instances of model construction result in what we call intermediate models. Together they provide small steps to construction of the target models. It is at this level that the instructor will employ multiple strategies such as analogies, discrepant questioning, discrepant events, and hands on activities to help students criticize and revise their current model. The instructor supports the process by introduction of new dissonance sources and by encouraging continuous application, criticism, and revision by the student. Figure 3.3 is an example from the *Energy in the Human Body Curriculum* of multiple intermediate models that were necessary to move a student from a simplistic model that all cells were just isolated, single cells (the fried egg model where students frequently draw an irregular shape with a circle in the middle) to the concept of cells as units of structure that make up the body. In this particular case it took three steps with multiple strategies in each step to help the student modify the initial model.

3.2.3 Determining the Intermediate Models Concepts to Use

Now that the need for developing a realistic set of intermediate models to target models to the final target has been established, it is up to the individual instructor to determine what the most effective targets for their unit of study are for his/her particular students. A target model (IMTS) is a mental model that students can effectively use to explain real life situations and one that is more scientifically correct than their initial models although it may not be absolutely correct or complete at the expert level (e.g. we may be justifiably satisfied at the elementary level with a Rutherford ("orbit") model of the atom or a monotomic model of gases). Using target models as the goal of instruction, students are encouraged to develop relatively complex models within their level of understanding and ability. However, how do instructors decide on the appropriate target concepts, as well as the target models and intermediate models so important to effective student model construction? Curriculum guides cannot possibly hope to address every preconception students will have or to develop a strategy for every conceivable circumstance. Therefore, it is important for the instructor to develop this path as they begin to plan for teaching.

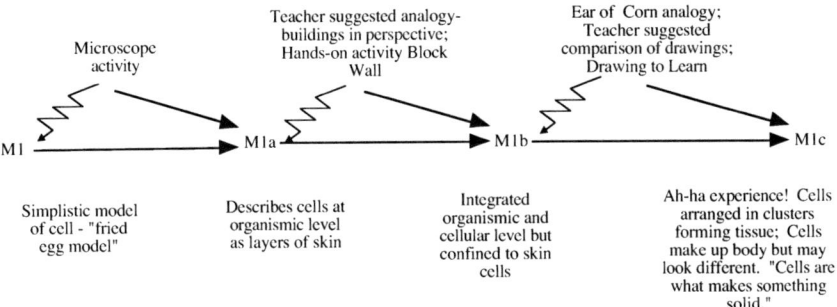

Fig. 3.3 Sequence of intermediate models, designated by M1a through M1c through which students moved in their attempt to criticize and revise their initial model of cells as structural units of the body. Multiple strategies were necessary to stimulate the criticism and revision cycles at each intermediate model stage

This determination of the final target model is often predetermined by State and Federal Science Standards, district curricular guides, and the adopted textbook. It is enhanced by an understanding of the current research in the topic area and on realistic targets based on the academic and developmental level of their students. It is important that the target be reasonable in its complexity so that it can be reached within the timeframe available. That is, it is not reasonable to assume that a target concept such as the Optimum Target Model (IMTS) for the Energy in the Human Body Curriculum,

> "Students will demonstrate an understanding of a dynamic model with both structure and function, and integration at the organismic and cellular level, as well as between systems. Shows causal relationships. Factors that affect energy need increase the need for oxygen and glucose, which in turn increases rate of breathing and heart rate,"
>
> could be attained in a short two weeks. This target is designed for a nine to twelve week unit. However, in two weeks it might be reasonable that students could construct a final target,
>
> "The circulatory system supplies the transport mechanism to supply oxygen and glucose to the cells of the body and is a closed system."

The instructor then identifies how realistically to break the final target into steps or target models. An attempt should be made to identify large concepts that build on one another and to keep in mind the importance of integration so that the final model that the students develop is integrated and dynamic, not merely a series of disconnected models that are not useful for explaining causal relationships. Looking again at Fig. 3.1, in the Energy Curriculum, the target models were developed to provide a pathway that builds on prior constructed understanding. That is, development of a circulatory system as a mechanism of transport for oxygen and glucose is necessitated by the need to explain how the elements arrive at the cell. In a previous target, students constructed understanding that these elements are needed by the cell to transform ADP to ATP. We call this necessity an Explanatory Need. Without this explanatory need, instructors and students could fall back into the traditional mode of learning isolated systems consisting primarily of structure and not function. When this occurs they see little connection between systems and therefore do not ultimately develop a dynamic, integrated target model.

Once a set of target models and the final target are delineated, only then can the instructor focus on designing reasonable strategies to aid the student in moving from one target to another. At this point the instructor identifies what preconceptions the students hold along with what the research says about students' misconceptions and difficulties in learning the topic, and then breaks the target models into intermediate models. The resulting series of intermediate models, target models, and final target model become the Learning Pathway.

3.2.4 Developing a Learning Pathway

Assisting students in moving from one target to the next rarely occurs in one large step. It takes carefully designed strategies and lessons to help students criticize their current model, make modifications, and test the new mental models. Having determined the final target concept, and then determined realistic target models, it is possible to outline steps that would provide a learning pathway consisting of multiple intermediate models. However, unlike the final target and target models, intermediate models may vary in response to the preconceptions students hold and to the specific needs of a group of students. This does not mean that the instructor can not determine some of these intermediate models prior to instruction but that the instructor must keep in mind that if in response to a strategy, the student appears, through gesture, posture, or verbal response, to develop too much or too little dissonance to prompt change, or begins to move far from the intended direction in model construction, then the instructor must make an "on the spot" decision whether to introduce another source of dissonance, to temporary terminate the strategy, or to try to further elaborate on the strategy. This often necessitates an instructor constructed intermediate model to bridge the gap.

In the Energy Curriculum, as in most instructor designed curricula, while the planned lessons or strategies are basically consistent, this does not preclude different reactions to the student's specific difficulties. To illustrate how a final target, target models, and intermediate models concepts are determined, we look again at the development of the Energy Curriculum (Fig. 3.4). Repeated modeling cycles were planned to help the students move from their initial model of respiration to a new, more scientific, target model (M6). A small segment of these (M1 to M2) cycles is presented linearly in Fig. 3.4 to allow for clearer presentation. However, rather than suggesting that a student can move from M1 to M2 in one cycle, there are multiple cycles of criticism and revision as students construct in incremental steps a more complete model.

Instructor - recognizes M1, generates M1a intermediate model.

Supplies dissonance strategy to stimulates construction of M1a, then M1b, M1c, until target M2 is reached.

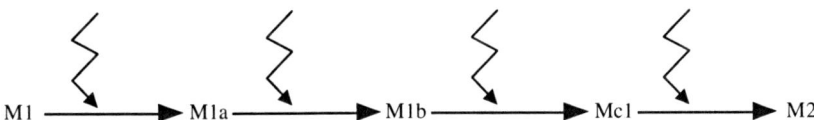

Fig. 3.4 Determination of Intermediate Models and generation of the next Intermediate Models along a learning pathway

This pathway is supported by the instructor or student generated intermediate models. These may be incomplete, partially correct, or naive models that allow the student to construct a new element of a model before engaging in further criticism and revision cycles. To understand just one concept may take many cycles to move from M1 to M2 and many, many more to move from M1 to M6. The zigzag lines indicate possible sources of dissonance while the straight arrows indicate movement in stages of construction over time.

3.3 Example of Development of a Learning Pathway in the Energy Curriculum

To promote student mental model development and reach the next target, a variety of strategies were used during each concept development step. This allowed students to move through repeated cycles of criticism and revision of their model. For example, to construct the target, "All cells have certain properties and functions in common, one of which is to produce ATP for the energy needs of the cell," required a series of intermediate models and a minimum of four strategies. The instructor first determined the students' preconceptions. A common preconception in this instance was the "fried egg" model of a cell. That is, students will initially draw their model of a human cell as an irregular circle with a smaller circle in the center. There is no indication of other structures and rarely any labels.

In response to the preconceptions identified, the instructor determined to help the students to construct a generic cell with organelles by first constructing an analogy of a school. This functional and structural model can then be mapped to the "fried egg" model of the cell to attempt to modify it gradually into a structure with more functioning internal structures. These are both considered intermediate models because, while they resemble properties of a cell, they do not yet include the complexity of function of the target. The next intermediate model would move the students toward giving function to the organelles and the general need for energy to carry out these functions. The next intermediate model would build on the previous model, but begin to focus on the mitochondria as the site for transformation of energy. This, in turn, would lead to the construction of a model of what the mitochondria needs to carry out this process and how it uses these products. Together the intermediate models gradually build in incremental units into the target concept, as depicted in Fig. 3.5.

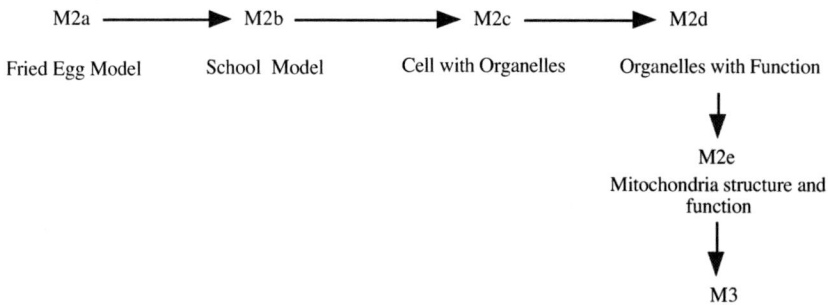

Fig. 3.5 Intermediate Models developed along a learning pathway from the students' preconception of a cell to the Target, M3

3.4 Determining Strategies to Stimulate Movement from One Intermediate Models to the Next

Once the basic pathway has been determined, the instructor can begin to identify strategies that will help in the criticism and revision cycles necessary to move students closer to the next target and eventually the final target model (see Fig. 3.1). Curriculum guides and the research literature are full of strategies that have proven beneficial in student constructed understanding. However, which strategy and when to use a particular strategy is really the domain of the instructor. While a curriculum can suggest what might be most effective, it is the instructor who must ultimately determine what is most effective and efficient for the time and topic. However, having a sound basis for that decision, grounded in research, will promote more effective model construction that simply relying on a collection of activities without thought to what intermediate model is to be constructed next and what the effect of the strategy is on the model construction process.

If, in the example in Fig. 3.5 a student had insisted that the cell could not have the structures of the analogy, then movement from M2b to M2c would not occur even though the instructor used the predetermined sequence of strategies. The instructor would then have to determine another strategy and possibly another intermediate model he/she could use to help the student to become dissatisfied with the current model and move closer to the new intermediate model. This might mean using a different analogy, like the traditional, "the cell is like a factory", or a hands on activity that shows an electron microscope photograph of cell where students could see there was "something else" within the boundary of the cell. Even though they could not necessarily identify the structures at this time, it would act as a dissonance producing strategy to stimulate dissatisfaction or criticism of the current "fried egg" cell. The instructor could then reintroduce the school analogy to attempt to construct the next model. The choice of strategies should be determined by the prior model students hold, how deeply held the model is, and evidence from the research that certain strategies have been found effective in helping students modify their model.

Therefore, in developing a learning pathway for a unit of study, the instructor needs to not only determine the final target and target models, but think about what small steps are realistic and reasonable to build the target. These small steps lead the student through model element criticism and revision cycles that lead from one intermediate models model to the next. It is important that this not merely be a of group of activities, vocabulary words, or lecture notes, but rather a well thought out plan for building on each successive model as it becomes more complex and dynamic. The ultimate goal is to develop a model that is "runnable." That is, an interactive mental video that the student can mentally "watch" from beginning to end as well as interjecting various constraints on the model and viewing causal relationships. An example might be a student who has constructed a dynamic model of digestion. In running the mental model, the student can see the process of food as it passed through the system, is digested, and absorbed into the circulatory system. He or she could also envision what would happen if a disease affected the villi in the intestines and caused them to flatten or decrease in size and number. From this runnable model the student could diagnose what the effect would be.

3.4.1 Suggested Procedure for Constructing a Curriculum

The complete curriculum would include the final target, target models, and intermediate models as well as the strategies that would be used to support criticism and revision cycles of the model. In order for learning to take place, a curriculum developer needs to understand both what preconceived mental models students bring with them to the classroom as well as what the research literature has already identified as important misconceptions and difficulties students experience in the topic area. Once this is established the instructor is ready to begin a series of steps toward development of a learning pathway that will hopefully support the student in constructing dynamic, integrated mental models that are "runnable" and applicable. These steps include:

1. Identify the final target concept based on Science Frameworks, National Standards, and district or school guidelines for the course.
2. Determine whether the target concept can be achieved within the time frame available. If the time cannot be adjusted, then the target will need to be reconsidered to make it reasonable.
3. Identify target models that represent major concepts that are inherent to achieving the final target concept. What pieces or model elements will the student need to construct in order to make the final target dynamic and integrated?
4. Determine how you will know when and whether students have mastered the target. What tools will you use to measure understanding? How will you visualize and document both for yourself as the instructor and for the student how the student's model has changed in the move toward the target? How will you measure integration between target models? Finally, how will you measure whether the final target

Determining Target Models and Effective Learning Pathways 57

is achieved and if not, what stumbling blocks, persistent misconceptions, or other difficulties exist?
5. By analyzing students' mental models at the beginning of the unit, determine what intermediate models may be needed to move students in small steps of model construction toward the target. Keep in mind that often even when very good intermediate models have been identified, student reaction to a certain strategy may move the construction process in a different direction than what the instructor planned. This often calls for the instructor to make 'on the spot' decisions about new intermediate models in order to help students refocus on the learning pathway.
6. Design a set of strategies that build on one another and aid students in constructing first the intermediate model and then the target. Strategies should be diverse to meet the various learning styles of students as well as providing multiple ways of supporting criticism and revision cycles. For more detailed descriptions of various strategies and how they provide support, see chapters in section II of this book.

3.5 Conclusion

This chapter has presented the reasons and strategies for developing learning pathways made up of a final target concept, target models and intermediate models. Research on students' preconceptions and possible learning pathways, has led to the design of a carefully structured sequence of activities for a curriculum, Energy in the Human Body, which has been used as an example of this process of developing a learning pathway (Rea-Ramirez, 1998). This research has provided a model for using target concepts to teach complex topics in science. The activities associated with movement toward the target concept are sequenced to develop and build on a series of intermediate models being constructed by the student in many small steps, taking care to have the student encounter small challenges that produced neither too little nor too much dissonance, and that could develop the student's model piece by piece in a manageable way. These intermediate models lead to the construction of the larger target models identified as important for understanding construction of a complex integrated target conception. This occurs primarily through incremental revisions in the models, with criticism and revision cycles. Careful identification of the final target and target models is important so that instructional goals are realistic and reachable. Furthermore, careful and deliberate identification of intermediate models is important to promote fective criticism and revision cycles. Monitoring of student responses to the strategies used to move students through a series of criticism and revision cycles from one intermediate models model to the next, is important to enable the instructor to modify or add details to the plan when necessary to promote learning and model construction.

References

Bishop, B. A., Roth, K. J., & Anderson, C. W. (1986). Respiration and photosynthesis: A teaching module ERIC document 272 382.
Camp, C., Clement, J., Brown, D., Gonzalez, K., Kudukey, J., Minstrell, J., Schultz, K., Steinberg, M., Venemen, V., Zietsman, A. (1994). Preconceptions in mechanics: Lessons dealing with conceptual difficulties Dubuque: Kendall Hunt.
Chiu, M. T. H., Chiu, M. H., & deLeeuw, N. (1991). Learning in non-physical science domain: the human circulatory system ERIC Document Reproduction Service No ED 342 629.
Clement, J. (1993). Using bridging analogies and anchoring intuitions to deal with students' preconceptions in physics. *Journal of Research in Science Teaching, 30(10)*, 1241–1257.
Gellert, E. (1962). Children's conceptions of the content and functions of the human body Genetic Psychology Monographs, 65, 337–345.
Nagy, M. H. (1953). Children's conceptions of some bodily functions. *Journal of Genetic Psychology, 83*, 199–216.
Niedderer, H. & Goldberg, F. (1996). Learning processes in electric circuits. Paper present at Annual meeting of the National Association for Research in Science Teaching.
Porter, C. S. (1974). Grade school children's perceptions of their internal body parts. *Nursing Research, 23(5)*, 384–391.
Rea-Ramirez, M. A. (1998). Models of Conceptual Understanding in Human Respiration and Strategies for Instruction (doctoral dissertation, University of Massachusetts, Amherst, 1998) Dissertation Abstracts International, 9909208.
Rea-Ramirez, M. A., Nunez-Oviedo, M. C., Clement, J., & Else, M. J. (2002) *Energy in the Human Body Curriculum*. University of Massachusetts, Amherst.
Sanders, M. (1993). Erroneous ideas about respiration: The instructor factor. *Journal of Research in Science Teaching, 30(8)*, 919–934.
Songer, C. J., & Mintzes, J. J. (1994). Understanding cellular respiration: An analysis of conceptual change in college biology. *Journal of Research in Science Teaching, 31(6)*, 621–637.
Tait, C. D., & Ascher, R. C. (1955). Inside-of-the-Body Test. *Psychosomatic Medicine, XVII(2)*, 139–148.

Chapter 4
Co-construction and Model Evolution in Chemistry

Samia Khan

University of British Columbia

4.1 Overview

Co-construction and model evolution is described in this chapter as a process of teachers guiding students to enrich their expressed models of a phenomenon. This process occurs when teachers ask students to explain relationships that are present within a student's expressed model. This chapter provides examples of how teachers can help students construct and express enriched models.

4.2 Introduction

Chemists model the physical and chemical properties of matter in an effort to explain why matter behaves in certain ways. For example, in the case of the structure of saturated cyclic compounds, the mechanical properties of a physical model and its inherent flexibilities and rigidities were closely linked to the development of hypotheses about the conformational behavior of these compounds (Francoeur, 2000). In this case, physical models not only acted as an aid to visualizing structure but also were used to explore the implications of the structure on the behavior of molecules. There are numerous examples of how the construction and evaluation of models in chemistry has influenced understanding of the nature and behavior of molecules.

A goal of teachers in chemistry is to promote student understanding of chemical interactions at the molecular level. For students of chemistry, tasks that require reasoning with their mental models may afford students with opportunities to explore the behavior of molecules and the nature of chemical interactions. But there are few strategies reported in the literature on how chemistry teachers can help their students construct, express, and reason with their mental representations.

This chapter provides examples of how teachers can guide their students to enrich their mental models by encouraging students to focus on important relationships that make up the model. Drawing upon a learning episode with a pair of introductory chemistry students and their chemistry teacher, this chapter reveals how students can construct relationships between 2 variables using lab information; how their relationships can become successively more complex to include 4 variables, and how students are able to reason through conceptually challenging material to eventually elucidate a molecular structure. His students and his peers identified the teacher selected for this study as an exemplary university teacher. The learning episode that is described in this chapter was part of a three-year study of student-teacher interactions in this teacher's undergraduate introductory chemistry course where the students were enrolled (Khan, 2002).

In contrast to a common public image of chemists at work with lab equipment, the work of analytical chemists has been described as that of handling mental representations, and that chemistry is to a great extent a "science of mind" (Francoeur, 2000). This chapter shares an analysis of how this exemplary teacher co-constructed mental representations with his students, and how these enriched models helped the students to explain chemistry phenomena. The chapter offers a description of student learning episodes and the teacher's strategies for model co-construction and evolution.

4.3 Model Co-construction

Chemistry teachers may have different goals and objectives depending on their pedagogy, curriculum, and students. However, it is virtually unanimous that all chemistry teachers wish to foster conceptual understanding of chemistry. In a teaching and learning episode that is the focus of this chapter, I investigate how the teacher attempted to foster conceptual understanding of molecules and the forces that exist between them, without directly telling students about bonds. Rather, the teacher's main strategy was to have students develop and express mental models of molecular structures that would help them explain observable chemistry, such as laboratory findings. Generally, model construction was encouraged by having students gradually build semi-quantitative relationships among variables that were central to the concept being taught (e.g. Whether an increase in variable *A* causes an increase or decrease in variable *B*). Students built semi-quantitative relationships by examining lab information and attempting to offer explanations for this information. Students inferred the properties of molecules to try and explain this information. In the end, students expressed more enriched models of molecular structures than those they had expressed initially.

This approach is different from a "teaching by telling" approach in that the goal is to guide students to construct their own models and use these models to explain laboratory findings. The approach involves the teacher encouraging students to express their mental models, guiding students progressively to enrich their model, and then asking students to test their model. Diagrammatically, the general set of conceptual exercises involves teacher and student engagement with the following sequence of activities:

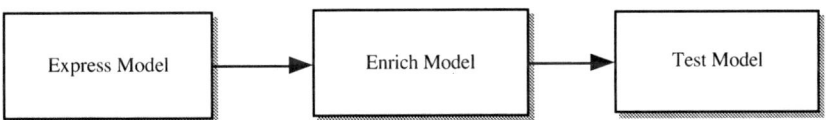

Fig. 4.1 Teacher-student engagement with model construction

First, asking students to make predictions about laboratory results and then explain the reason for their predictions accesses students' prior knowledge. This activity allows students to express their thinking about chemistry and reasoning at a molecular level. For the teacher, this activity presents a potential opportunity to inspect students' understanding of relationships in chemistry and their prior mental models of molecular structures. Once these mental models are revealed, the teacher is in a position to work with the student to enrich their mental model. Secondly, the chemistry teacher seeks to enrich students' mental models by co-constructing connections to more variables. Indicators that a student's expressed model appears to be more enriched include: (a) an addition of new variables to the model, (b) an addition of modifiers to relationships within the model, (c) use of relationships within the model to explain a novel lab finding, (d) change of drawings of molecular structures, and (e) ability to make predictions and offer explanations that students were not able to do so prior to the learning episode. Finally, students' expressed models appear to evolve significantly when their models are tested. The final stage of the instructional sequence involves the chemistry teacher asking students to explain lab information that provokes them to reconsider their original models of molecular structures. Following the instructional sequence, the next sections illustrate specific teaching strategies that can uncover students' initial models, enrich student models, and engage students in testing their models.

4.3.1 Expressing Models: Accessing Students' Prior Knowledge with Surveys

In order for teachers to construct a model with their students, they can begin by building upon what students already know about the phenomenon in chemistry. Students' knowledge prior to the teaching interaction may be scientifically accurate already, or, alternatively, students' may have limited experience with the content and partial understanding of the concept. In the case that I draw upon for this chapter, a student pair was surveyed with open-ended questions for their understanding of molecules and their effect on vaporization prior to the teaching episodes presented here. The survey gauged students' understanding of the following six conceptual relationships in chemistry:

1. Increasing temperature increases molecular speed.
2. The greater molecular mass, the lower the average molecular speed.
3. The lower the molecular mass, the higher the vapor pressure.
4. The more H-bonding in a compound, the lower the vapor pressure.
5. The greater the molecular mass, the higher the boiling point.
6. The more H-bonding, the higher the boiling point.

The initial survey revealed that both students understood the relationship between molecular mass and temperature on molecular speed prior to the teaching episode but that neither student understood how the structure of the molecule affects vapor pressure or boiling point. For example, in response to survey item #6, one of the students (S1) explained that a hydrocarbon chain would have a higher boiling point compared to a hydrocarbon chain with a hydroxyl attached. The reason was that, "There are more molecules to separate &/or excite, so it will take longer and more temperature to do so." Thus, findings from this survey revealed that the students had only a partial understanding of properties of molecular structures and molecular behavior under different conditions. For teachers, surveying their students can be a strategy to gain access to their prior knowledge.

4.3.2 Enrichment of Expressed Models: Prediction-making Activities Based on Prior Models

In order to encourage students to reveal their models further, prediction-making activities are suggested. The teacher asks students to make a prediction between two specified variables. Asking students to make a prediction affords students with the opportunity to express their initial models to the teacher for inspection, and focuses students on the activity ahead. In the excerpt below, we see how two students (S1 and S2) respond after their teacher asks them to make a prediction about the effect of raising the temperature on the behavior of molecules. The teacher asks students to make this prediction based on a graph that showed a Boltzmann distribution of molecular speeds at specific temperatures.

> S1: I was going to say, it [a Boltzmann Distribution] definitely would move to the right, but I suppose it would flatten out, too.
> S2: Because there are still molecules that are going to be going real slow.
> S1: So, the average would be more dispersed?
> S2: Yeah, and there's the same number of molecules altogether.
> S1: Molecules move faster, I believe.

This excerpt shows that students consider average molecular speeds and recognize that some molecules will be traveling above or below this speed at a given temperature. Students then propose a mechanism for the increase in molecular speed that involves energy and excited particles, in response to the teacher (T) asking for an explanation for why molecules move faster when they are hotter:

> T: And why do they move faster if they're hotter?
> S2: They have more energy?
> S1: Yeah.
> T: Okay.
> S1: They're more excited.
> T: They're more excited, they have more energy, they move more quickly. So, in general, you would say, 'As temperature goes up...' What's the general rule between temperature and molecular speed?
> S1: The hotter the temperature, the faster the molecular speed – the higher the temperature.

The students infer that molecules become more excited with an increase in heat energy and this plays a role in moving them faster. Students express to the teacher, in this prediction-making activity, a molecular model that involves a relationship among molecules, energy, motion, and temperature. It is important to note that after the teacher asks students to make a prediction and when students offer one, the teacher follows up with a request for an explanation. Asking students "why" appeared to encourage students to make inferences at a molecular level. The outcome of this episode is that students made a prediction, suggested a molecular level explanation, and constructed and expressed an initial relationship among variables; that is, the hotter the temperature, the faster the molecular speed. Taken collectively, asking students to make predictions and offer an explanation in this learning episode appears to afford students with a way to express their model of how molecules behave under different conditions.

4.3.3 Enrichment of Expressed Models: Building on Relationships in a Sequence and Drawing

The teacher then asks students to build on the first relationship regarding the speed of molecules, but this time, between two different variables: molecular weight and speed of molecules. Again, he asks students to make a prediction and provide an explanation for the prediction. Again, we see students express a model of molecular behavior in an attempt to explain their predictions.

> T: I want you to predict what you think it [Boltzmann distribution] will look like for helium and xenon, both in comparison to oxygen at 600 Kelvin. So, draw a little box like that for where you think those will be. And talk about why that is-why you predict what you predict.
> S1: If you have some water-a little bit of water-it doesn't take as long to heat up as like a big cup of water would take to heat up, and the xenon weighs more-there's more mass in it, so I don't know if that...
> S2: For the one that's heavier, it would take more energy to get it to be going at the same speed as something that's lighter.
> T: Right. So, to get the xenon curve to look like the oxygen curve, you need to do what?
> S2: Heat it up.

The student pair has constructed the sequence that as temperature increases, the speed of molecules increases, and now, as mass of molecules increases, the speed of molecules decreases. The teacher continues to encourage students to explain these relationships, and this strategy appears to provoke explanations by students of the behavior of molecules.

This next learning episode results in enriched student models. As the prior survey indicates, students were not able to explain how vapor pressure affects molecules. We see an apparent change in students' explanations as they interact with their teacher.

The teacher begins with the definition of vapor pressure and then asks students to make a prediction about vapor pressure as temperature increases. Once students make a prediction, the teacher asks students to explain what is happening at the molecular level by suggesting that students draw what they are envisioning.

> T: What do you think will happen to the vapor pressure as temperature is increased? Okay? So predict the plot. So there are two things you're predicting: what's the general direction? As temperature goes up, what happens to the vapor pressure-up, down, stays the same? And then is it linear, curved, what do you get?
> S1: Well, do you think that as temperature goes up, vapor pressure goes up?
> S2: Mmm, yeah.
> S1: I can't explain why I think this. I don't think that it [the curve of the Boltzmann distribution] would just grow exponentially and

just go off forever. I don't know. And it has to be at a certain point in my mind when it would start peeling off just a little bit. It'll always continue to grow, but not at such a sharp incline. [S1 spontaneously draws graph below of vapor pressure versus temperature].
S2: Probably.

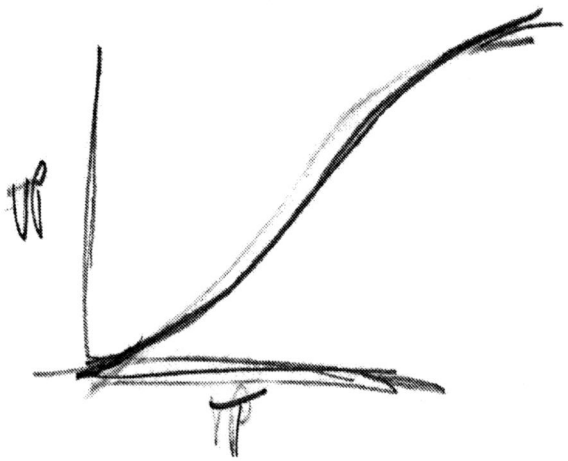

Fig. 4.2 S1 drawing of a graph of vapor pressure versus temperature

T: So, let me ask you about this. You both say it's going to go up with temperature?
S1: Well, since things move faster-molecules move faster as temperature goes up...
T: Draw pictures of how you envision that happening-that stuff going from liquid to gas on the molecular scale. So like draw pictures of molecules of methanol-and draw what does it look like in the liquid and then what does it look like in the gas?

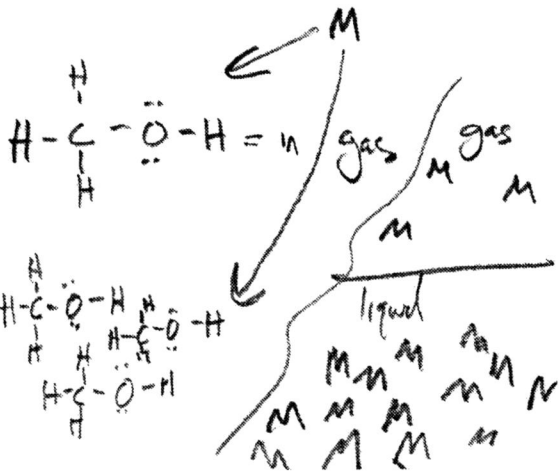

Fig. 4.3 S2 molecular level drawing of methanol vaporizing

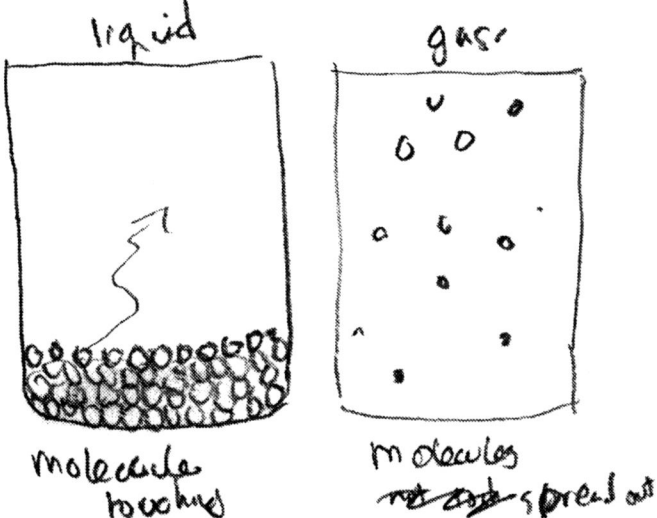

Fig. 4.4 S1 molecular level drawing of methanol vaporizing

S1: We don't think there's any structural difference, I guess [pointing to the drawings]. It's just the amount that's found in a certain area, and the gas that's far more spread out.
T: Okay.
S1: In the liquid form, they're condensed-close together.

> T: Okay. In fact, both of you-your intuition is correct you've described this very well. In the liquid phase, you've got all these molecules that are near each other, in the gas phase they're the same things, but more separate from each other.

In the excerpt below, students go on to suggest that the molecules gain enough speed to "break out of the liquid", to "break apart" and enter the gas phase when the temperature increases:

> S2: Okay. Um, when the temperature increases, there are more molecules that are going at a-I made this line here [refers to Fig. 4.3. drawing], when I was thinking about this- there are more molecules that are going faster.
> S1: Yup.
> S2: And, they have enough speed to *break out of the liquid*.
> S1: *Break apart* from each other.
> S2: And, then sometime-and like, yeah, more of them will stay in the gas when it's at a higher temperature, because there will be a higher vapor pressure, I guess.

In the above two excerpts, we see how students' models can be inspected by the teacher through drawings. In this case, students are asked to draw models of molecules in the liquid and gas phases. Students' drawings reveal particles touching one another and that the particles are structurally similar in the liquid and gas phase. The apparent difference between molecules in gas phase and liquid phase is how close the particles are to one another. Students point and refer to their drawings as they begin to explain the mechanism of how molecules actually enter the gas phase. (In this episode, an interviewer is also present with the teacher and student pair as denoted by I in the following transcript). Italicized portions of the transcript are referred to later in this chapter.

> S2: [The molecules are] separate and they bump into each other, and *they might stick together for a short time*, but mostly the molecules are separate.
> S1: Not right next to each other. They still have to have room to move. If they're right next to each other, or even-the way I picture it is solids don't necessarily have to be touching. There just has to be such little amount of space between them that they have nowhere else really to go, and that's why they're packed in. With the liquid, there's much less room for-they can't just float around like this, but they have a lot more room to move than a solid would…

S2: Um. I think that-I still think that they're touching.

S2: No, they're separate and they bump into each other, and *they might stick together* for a short time, but mostly the molecules are separate.

S1: I don't think I would have used the word break [in reference to the italicized portion of the above transcript]. Now that I'm thinking about it, break isn't appropriate from my train of thought, because that would imply that they're actually connected. I'd more call it-*separation* would be better, or moving away from, because breaking implies that they were actually physically connected, and *I don't ever envision it as they're being physically connected.*

I: And what separates?

S1: The actual molecule itself. If you want to picture this as a group of balls, you know, it would look something like this. And you can picture a bunch of these snowmen-looking guys all *next to each other* like this-maybe a little bit more room. They've got room to move around each other, but then, when it heats up they're moving faster, and because they're moving faster, there's more opportunity for one of them to – you know, say there's a bunch of them-a lot more opportunity for one of them to completely make a-I don't want to say break, but I guess that's the word-to move off into open-I don't know, what's up here-oxygen, nitrogen, stuff like that.

S2 appears to suggest that there might be something that "sticks" the molecules together for a period of time, and S1 wishes to clarify that the particles are not physically connected, so when they enter the gas phase, the particles undergo more of a separation rather than a break (a word that he had used earlier). Here we see students' proposing new unobservable dimensions of their model such as "closeness", "stickiness" and "separations" among molecules in an effort to elaborate on molecular behavior under rising temperature certain conditions. Missing in their mechanistic explanations, however, is mention of collisions among molecules. The teacher proposes an idea to explain why there are more molecules entering the gas phase as temperature increases:

T: Let me ask a specific thing. If a molecule-one thing that could be happening-one possible explanation is there's some probability of the molecules getting away from each other, and if it's hotter, they're moving faster and therefore more often that will happen, just statistically because they're moving faster, and say 10% of the time-10% of collisions between molecules result in one going off-if they move faster they'll collide 20% more often, and therefore two more molecules will go away.

By the culmination of this interaction with their teacher, students have co-constructed relationships among several variables. The drawing activities appear to afford students with an opportunity to envision and elaborate on the structural and spatial dimensions of their models. As well, students spontaneously refer to their drawings to explain interactions involving molecules during changes in temperature conditions. These explanations sometimes involve students' postulating unique, unobservable dimensions of their model that were not expressed before. The student pair has thus far constructed the relationships that: as temperature increases, the speed of molecules increases; as mass of molecules increases, the speed of molecules decreases[1], and as temperature increases, vapor pressure increases. Using a similar process of co-construction involving prediction making, asking students for explanations, building on relationships and drawing, students also add the relationship that as molecular mass increases, "vaporization goes up". Adding variables to one's models provides some evidence that the model has been enriched.

In summary, building relationships sequentially, asking students to explain lab findings and draw what they are envisioning is happening at the molecular level, are three strategies that appear to contribute to students' further construction of their prior models of molecular structures.

4.4 Model Evolution Through Testing

4.4.1 Testing Established Models Through Comparison and Examination of Data

The final conceptual exercise in the instructional sequence has students testing their models under different conditions. In this phase, students' once again are encouraged to keep expressing their models of molecular structures. Change to expressed models becomes apparent, for example, with the addition of new variables, addition of modifiers to relationships, use of relationships to explain a novel lab finding, change of drawings of molecular structures, or ability to make predictions and offer explanations

---

[1] This section of transcript not included; however, the interaction is similar to the other sections of transcript shown.

that they were not able to do so before. Testing their models under different conditions appears to reinforce pre-existing models or in some cases, motivate students' to alter already expressed, established models. For example, in the following learning episode, students are asked to compare the vapor pressures of two compounds of different molecular weights (methanol and ethanol) using their previously established model of molecules.

> S1: We already established that the smaller the molecule is, the faster it will move, so the faster it moves, the more opportunity it has, the more chances of it escaping come up within a given amount of time, than would happen with the ethanol... We already established that the smaller a molecule, is the faster it will move when it gets heated up, proportional to a larger molecule heated at that same temperature. And, since we also established here that the faster a molecule moves, the higher its vaporization temperature will be, it only makes sense that the smaller molecule will have a higher vaporization.

S1 suggests that a smaller molecule would move faster than a larger molecule and therefore, have more of an opportunity to escape, resulting in a higher vaporization. S1 then tests this hypothesis against actual data available on the rates of vaporization of two compounds of different molecular weight, further confirming, in his view, that smaller molecules will have higher vaporization. Testing using this vaporization scenario reinforced the students' previously established understanding of molecular structures. Students are then asked to make a prediction about molecular weight and boiling point:

> S2: So, at a certain temperature, the molecular weight increases, and the speed of the molecules will decrease, because it takes more energy to get moving at that speed, so they're moving slower because they're heavier...
> S1: ... denser...
> S2: Because they're heavier... They're moving slower so they're less likely to break away-to go from the liquid phase to the gas phase than a lighter molecule, and so the boiling point will be at a higher temperature. It will be at a higher temperature before... the vapor pressure is one atmosphere, before it gets that high.

Students predict that higher molecular weight compounds will have higher boiling points than lower molecular weight compounds. They explain this relationship both at a molecular level (see transcript above describing behavior of molecules) and show the relationship in a jointly constructed graph (see Fig. 4.5).

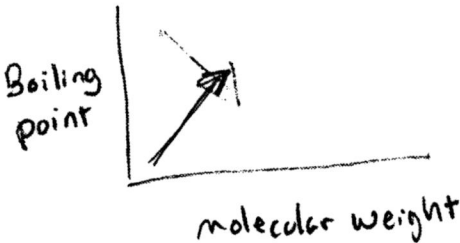

Fig. 4.5 Students' graph of boiling point vs. molecular weight

In a new testing scenario, the teacher asks students to predict what would happen if instead of a hydrocarbon chain such as ethane, they were to substitute a hydrogen with a fluorine in the compound:

> S1: When you add a fluorine, like she said, it's the most electronegative element there is, so just because you make your molecule polar-I don't know if that necessarily changes the fact, changes anything with regards to boiling point, because I think boiling point is more a function of mass than of polarity.
> S2: If you make it polar, though, what you're going to get is *interactions between the partial negative and partial positive*, and that's going to be interaction that's going to happen in the liquid, and if we go into the gas... it's not really bond, but you'd have to break *that attraction*.
> S1: So, what you're saying is it would be much more difficult...
> S2: ... so I think the boiling point would be much higher.
> S1: ... but it would still go in that upward trend.
> S2: Yeah.

S2 considers the presence of some form of attraction between partial positive and negative ends of compounds containing fluorine that needs to be overcome. This factor, a "force of attraction", was not mentioned by either student in the initial survey nor was it a topic that had been covered in the introductory chemistry course. S1 still asserts using the graph that molecular weight, rather than polarity, is the most important factor.

S1: Well, the molecular weight is still the biggest function-the biggest factor here. The fact that we have the partial positive and partial negative sides just means that your base point on this graph [referring to Fig. 4.5, graph of molecular weight vs. boiling point drawn above] is going to be higher than what it is right now, but it's still going to move in this upward trend.

Students then proceed to add the partial positive and negative charges to their drawn models of molecular structures. These charges were not evident in their earlier drawings (see Figs. 4.3 and 4.4).

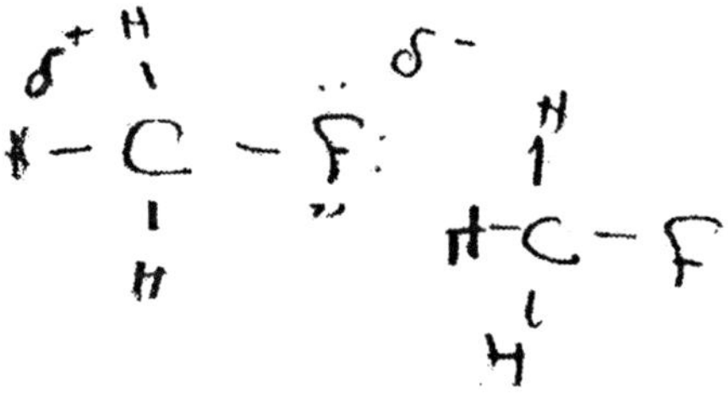

Fig. 4.6 Evidence of charges on student's drawings of molecular models

Students are then asked to test their ideas about molecular weight and polarity against recorded values of boiling points for different chemical compounds that were available from large data sets (e.g. methylamine CH_3NH_2; methanol CH_3OH, methyl fluoride CH_3F). S1 finds that his molecular weight theory is unsupported:

S1: Well, originally I thought that molecular weight was the overriding factor. The more something-the higher the molecular weight, the higher the boiling point was going to be, but we just found out that the amine is-actually weighs less than fluorine, yet it has a higher boiling point. Plus, the amine is-I think it's-it looks non-polar to me, which would go against the idea that something would have a higher boiling point if it was polar, which we thought because the fluorine's polar. So it kind of goes against both of our previous ideas

S1: Weight doesn't seem to have a correlation here, because the fluorine weighs the most and it's on the bottom [i.e. has the lowest boiling point compared to methanol and methylamine], which goes against what we thought before, or I thought before. *OH* is in the middle, yet it's on top, and our *NH₂*, which weighs the least, and which I figured would be on the bottom, is in the middle.

They redraw their graph:

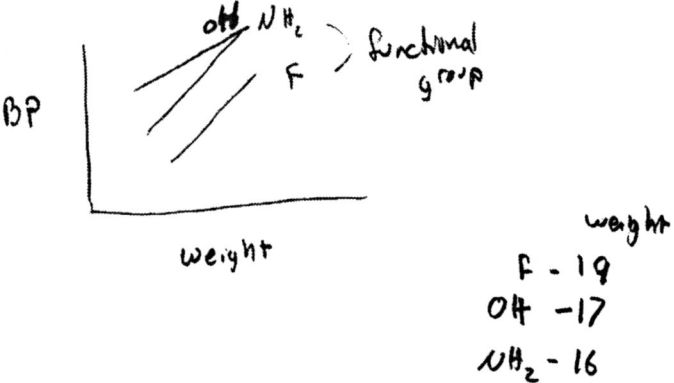

Fig. 4.7 Students' redrawn graph of boiling point versus molecular weight

Testing models against large data sets of lab information and comparing values of different substances appears to motivate students to examine and reconsider their original models of molecular structures. The students refer back to their drawing to reflect on its accuracy.

> S2: I guess if you have an attraction between molecules that needs to be overcome for them to break apart... that takes more energy to do that than just to get them going. [See drawing by S2 in Fig. 4.8].
> S1: Because...
> S2: It might be that this end hydrogen might be more partially positive than this whole thing would be from being attached to the fluorine.

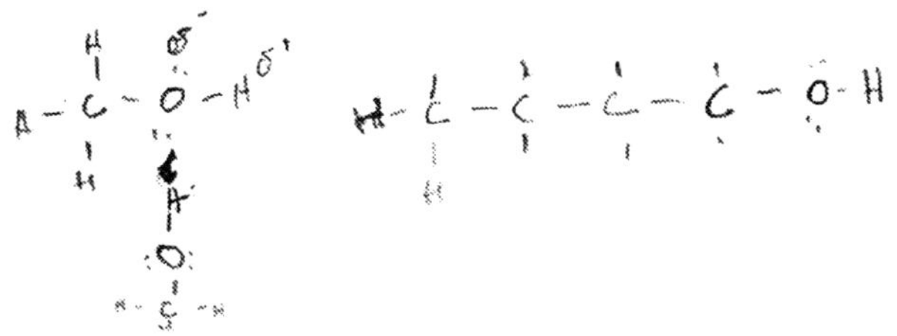

[Hand-drawn chemical structures showing molecules with partial charges]

Fig. 4.8 S2 drawing symbolizing 'attraction' between two molecules

T: Okay, that's actually- that's it. The fact that this oxygen-so you nailed it exactly-the fact that this oxygen or this oxygen has a really high electronegativity makes the hydrogen really positive, and then can be attracted to the other oxygen, whereas when you have a hydrogen attached to a carbon... it doesn't get very positive.

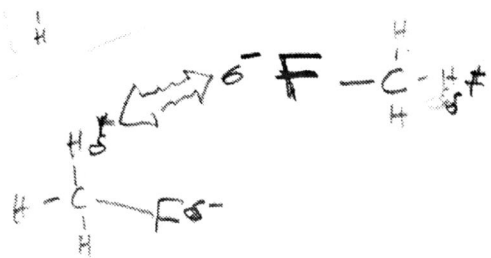

Fig. 4.9 S1 drawing symbolizing attraction between two molecules

T: So this can't get very positive, even though the fluorine here-- this second electron did stay away from that carbon, it doesn't do it enough to make it that positive.
S1: So this drawing [pointing to Fig. 4.9] in general isn't really right?
T: No, it's right.
S1: But it's [force of attraction] just not as strong as this one [pointing to Fig. 4.8].
T: Yes. So what happens is these kinds of interactions aren't a whole lot stronger than the interactions we just had with the hydrogen or carbon here. So these interactions end up being pretty similar to standard interactions that you just get for all things. And in that case, the thing you saw with molecular weight is pretty

much the rule. Whereas when you have hydrogens attached to really electronegative things, you get these really highly positively charged hydrogens and that leads to really strong interactions [pointing to Fig. 4.8].

Following this interaction with their teacher, the students are able to use their revised models that include polarity to explain new lab findings they had not encountered previously in the learning episode. For example, when asked to predict the boiling points of water in comparison with benzene, students predict that water would have a higher boiling point because of its polarity compared with benzene:

S1: So that would explain why it [water] takes-why it [water] has a higher boiling point [than benzene].
S2: Boiling point depends on the molecular weight and it depends on the polarity of the molecules or the interactions between molecules.
S1: It seems like polarity has a stronger effect than weight.

In the above passages the teacher asked students to examine large data sets of information on boiling points. Testing models against large data sets of information, such as data sets on vapor pressures and boiling points, appears to afford students with opportunities to propose new factors and re-consider their constructed models. In this episode, we see that the students' molecular model evolved to include polarity and a force of attraction between molecules, and a revised molecular model was used to explain novel lab information regarding benzene and water's boiling points. On the post-survey, the student pair progressed conceptually and performed better than in the initial survey on three of the six conceptual relationships they missed on the initial survey. In all, a sample of 12 students from the course were surveyed after a similar instructional experience; they too progressed significantly on the post survey compared with the pre-survey (paired t test, $p<0.001$; n = 12). Thus, we have evidence that students' new models of molecular structures consisted of connections that were not present before, such as the connection between molecular structure and boiling point, and that students' models appeared to have evolved compared to a partial model that was evident in the initial survey.

4.5 Summary of Model Co-construction and Evolution Learning Episodes

Chemistry is replete with models. This chapter uses a learning episode to illustrate how a chemistry teacher co-constructed with students a model of molecules that could explain vaporization and boiling. The teacher encouraged the students to express their model of molecules by focusing on single relationships within a model one at a time. The teacher did this by asking students to make predictions of the relationship between two variables and then explain the reason for their predictions. His request that they use drawings to explain appeared to be important. The teacher did not correct students' predictions, but in one case, he added some content to their explanations. Students constructed relationships and successively added variables to their models, resulting in models that appeared to be enriched compared with what they expressed earlier in an initial survey. The teacher then asked the students to test their model by comparing lab information on new compounds. Testing predictions to ascertain if they were accurate according to available data appeared to motivate students to consider new factors that they had not expressed before. Students' ideas about molecular structures appeared to evolve especially when students were asked to run a test under two hypothetical scenarios: vaporization and boiling point scenarios that compared substances using data sets. When their models were run under these conditions, students were observed reasoning how the structure of molecules and their interaction with each other could explain the data. This episode reveals how a pair of students was able to express, enrich and evolve their models in chemistry with questions and activities from their teacher. With the teacher providing a guiding framework of prediction and explanation questions, and occasional content input, the students were able to generate the vast majority of correct relationships in their models themselves. The teacher refrained from telling students the answers, yet played an important role in engaging them in a series of conceptual prediction and explanation exercises to develop their understanding of chemistry at the molecular level. This model of guided constructivism illustrates one form of teacher-student co-construction that was possible in a college chemistry classroom setting.

References

Francoeur, E. (2000). Beyond dematerialization and inscription: Does the materiality of molecular models really matter? *International Journal for Philosophy of Chemistry 6,* 63–84

Khan, S. (2002). Teaching chemistry using guided discovery and an interactive computer tool. (pp. 1–337). *Unpublished doctoral dissertation*, University of Massachusetts, Amherst.

Chapter 5
Target Model Sequence and Critical Learning Pathway for an Electricity Curriculum Based on Model Evolution

Melvin S. Steinberg
Smith College

5.1 Introduction

This chapter discusses the design of a recently developed high school electricity curriculum called the Capacitor-Aided System for Teaching and Learning Electricity (CASTLE), which aspires to enable students to construct a sequence of increasingly complex qualitative models of electric circuits (Steinberg & Wainwright, 1993). The curriculum is driven by hands-on student experiments on bulb lighting in circuits that contain batteries and capacitors, sequenced to foster a learning pathway of model modifications that add conceptual complexity gradually with low cognitive load. The growing complexity periodically requires a revised conception of causal agency. The transition from emission by a battery to pressure in a compressible fluid as the agent of current propulsion is described here. Transitions to increasingly abstract causal agents of distant action will be described in a later chapter in this volume. Teacher and student manuals with complete details are available in *Electricity Visualized* (Steinberg et al., 2007).

5.2 Principles of Curriculum Design

CASTLE curriculum design principles were mostly tacit while the development team was identifying target models and learning pathway.

Classroom experience, workshops for teachers, and student interviews have since made it possible to articulate the following principles.

P-1: Anchoring conception

The model building sequence begins with a widely held valid preconception (an anchor in our terms). For electric circuits, this is the belief that "something is moving in the wires" connecting a battery to glowing light bulbs, which in the experience of the development team is universally held by beginning students.

P-2: Conceptual dissatisfaction

Surprising observations, especially those that conflict with preconceptions, are used to create a need for model modification. When such "discrepant events" are not available, teachers may ask "discrepant questions" that provoke dissatisfaction with the existing student model.

P-3: Observational constraints

Hands-on experiments are used to provide observations that constrain model building in a productive direction by

- suggesting important new concepts
- disconfirming alternative conceptions
- separating conflated science concepts

The CASTLE electricity curriculum uses novel bulb lighting experiments for this purpose.

P-4: Gradual model modification

Fostering model modification in small steps can make complex model building doable for beginning students. The advantages over attempting to foster large conceptual leaps are:

- Most of the prior student model M_i remains intact and available to support reasoning during each episode of conceptual change.
- A failure of the newly perceived flaw in the existing model can be corrected without negative impact on confidence when only a small modification is needed.
- Assuming M_i is a runnable model and makes sense, then M_{i+1} should be runnable and make sense if the change was not too great.

These conditions are designed to sustain students' confidence and interest in active learning.

P-5: Representation in dynamic imagery

Experimental investigations are chosen for their ability to stimulate manageable imagistic reasoning. The goal of the experiments is not measurement and confirmation of a principle, but enabling students to run mental simulations that reveal consequences of their existing model and of proposed model modifications. Using dynamic imagery to represent physical relationships has advantages for beginners over mathematical representation favored in conventional courses:

- Multiple constraints in a physical system can be represented in the same image.
- Imagistic relationships allow analysis by means of concrete spatial reasoning.
- Imagistic simulations can evaluate model modifications and solve novel problems.

P-6: Transfer of dynamic imagery

Each surprising observation is explainable by a target domain process that students regard as similar to an analog domain process where they can make confident predictions. The analog domain is selected for transferability of the dynamic imagery associated with it into the target domain. This transfer can be especially powerful for kinesthetic imagery associated with causal agency. Clement has studied expert subjects for whom analog models of mechanism generated by kinesthetic-visual experiences embody dynamic imagery that is transferable (Clement, 1994, 2003, 2005, 2008) and for whom this imagery is transferable (Clement, 2003). Clement and Steinberg (2002) have studied a teen-age student who drew on dynamic imagery generated by concrete experiences with *air* to conclude that what moves in electric circuits is *always present in wires* and can be *compressed* to create "pressure" that drives electric current. (Clement & Steinberg, 2002) They hypothesized that this involved a transfer of dynamic imagery from the source analog to the target domain.

P-7: Anchor development

Although anchoring cases by definition utilize widely held valid preconceptions, students sometimes need help in developing an anchoring conception. Intuition of analog domain processes is strengthened by activities that fill gaps in imagery for the base of an analogy. The CASTLE curriculum uses balloons, air syringes and "air capacitors" to strengthen and complete students' imagery of air compression, pressure changes, and pressure-driven movement in air.

P-8: Imagery enhancement

Diagrammatic conventions are invented to provide external visual support for reasoning about changes of variables. This increases the level of detail in a visualizable model and makes it more powerful. The CASTLE curriculum adds three symbols to schematic circuit diagrams:

- color-coding of circuit wires to represent pressure magnitude in each wire
- starbursts drawn in the space around all bulbs to represent their brightness
- arrowtails drawn near each bulb to represented flow rates through the bulbs

It is important that these symbols be simple and schematic enough to support *mental imagistic* representations as well as those on paper. Changes of these symbols in successive time intervals can represent frames of a mental stepwise animation that helps students visualize a dynamic process with coarse-grained quantification. (See Fig. 5.9c, 9d, 9e and 9f below.)

5.2.1 Overall Approach

Classroom experience, plus research on reasoning by individual students and experts, suggests a strategy that helps students obtain imagery needed to run complex simulations in an unfamiliar domain:

- Find an analog domain where students already have a runnable imagistic model.
- Strengthen the analog model, if needed, with additional empirical experiences.
- Foster discussion of similarities and differences of imagery in target and analog domains.

Then elaborate the target model further as needed, possibly eventually leaving the initial analogy behind. Enhance imagery as needed with clear schematic diagrams and visual representations. Strive for maximum simplicity while retaining explanatory viability.

5.2.2 Critical Learning Pathway

An effective curriculum must help students overcome misconceptions that block learning. In electricity, the widely held and deeply believed preconception that a battery sends out what's moving in the wires connecting it to lit bulbs is a potent obstacle to learning (Clement & Steinberg, 2003). The need to provide experiences that show such conceptions are untenable and that suggest superior replacement conceptions is a constraint on effective curriculum design that critically directs the succession of model building steps. Once most of the useful steps have been found by talking to students and determining the sub-problems in building the target models, one has the outline of an effective learning pathway.

Articulation of a number of the above principles was inspired by think-aloud studies of expert learning processes. For example, see: **P-5**, **P-6**, **P-8**, Clement (2003); **P-4**, Niedderer and Goldberg (1996); Clement (1989, 2008).

5.3 Description of Curriculum Sequence

5.3.1 Initial Battery Emission Model

Circuits of batteries and bulbs alone suggest to beginning students a mechanism of bulb lighting in which *a battery sends something out through the wires* that connect it to glowing bulbs. In this conception the battery is (a) the only source of what's moving in the wires and (b) the only causal agent that can make that movement occur. The useful modifications that are possible within the scope of this battery emission model include circuital direction of movement and control of flow rate by resistance. These modifications involve so little added complexity that students in 9th grade can achieve them by conceptual accretion prompted by empirical observations.

M-1: The "circuit" idea

Students are now asked to connect three wires to a battery and two bulbs in a way that makes both bulbs light. Their finding is depicted in Fig. 5.1. The term "circuit" – derived from "circle" – is used to describe the wire connections that enable this dual bulb lighting to occur. Discovering these connections leads students to adopt a model in which bulbs glow only when wires connect them to a battery in an unbroken circuit. (Uses **P-3** and **P-4**.)

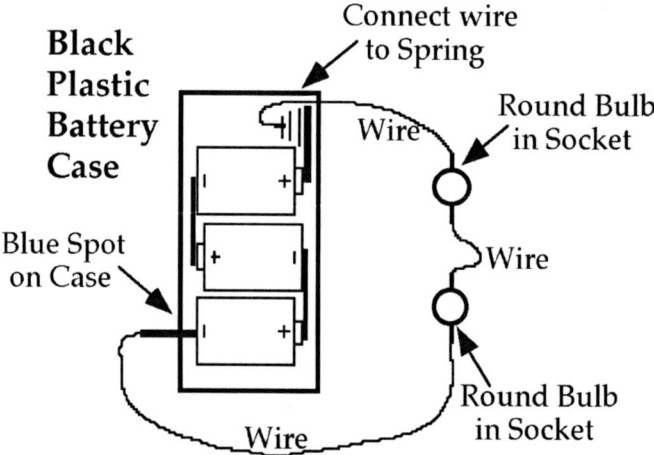

Fig. 5.1 3-cell battery and 2 bulbs connected by wires

M-2: Circuital movement

One student holds a wire down on top of a magnetic compass, parallel to the compass needle, while a partner closes the circuit and both observe the resulting needle deflection. When the battery polarity is reversed, the direction of the needle deflection reverses. Students regard this as evidence for reversal of *direction of movement* in the wire. They then regard the direction of needle deflection as indicating direction of flow, and the magnitude of deflection as a measure of flow rate magnitude. The under-wire compass is regarded as a device that looks at a circuit from the outside, and does not interfere with what happens inside the circuit. An ammeter is not used because it modifies the circuit being investigated. (Using a compass to detect magnetic field near a circuit with charge flow is postponed until late in the curriculum.)

In the experiment illustrated in Fig. 5.2, the entire circuit is rotated so as to place each of the three connecting wires sequentially on top of a fixed compass, which is taped down to avoid directional confusion that may occur if it is moved from wire to wire. Students observe the compass needle deflecting in the same direction for each of the wires placed on top of it, and they regard this as evidence that movement in the wires is either clockwise or counter-clockwise in all three connecting wires. Their depictive hand motions over the circuit suggest that observation of unidirectional compass deflections during bulb lighting generates imagery of unidirectional movement. What's moving in the circuit is given the name "charge". (Uses **P-1**, **P-3**, **P-4** and **P-5**.)

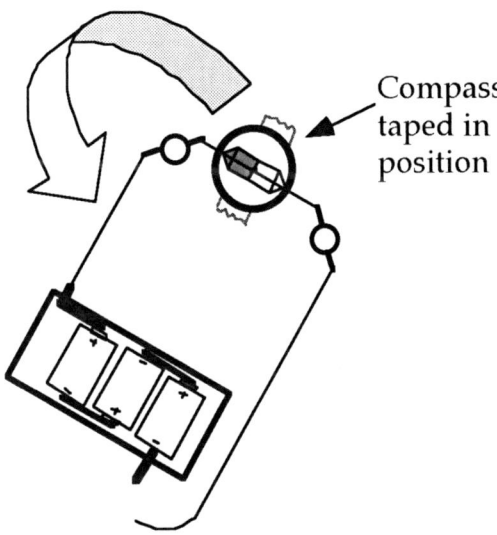

Fig. 5.2 Circuit rotated over a compass taped to table

M-3: Conductors and insulators

The "something to be tested" in Fig. 5.3 is a site where students may place various objects or materials in the circuit between the two bulbs. Observing whether the bulbs do or do not glow provides evidence that an object/material permits or prevents movement of charge in the circuit. This experiment allows an object or a material to be classified as a "conductor" that allows movement of charge through itself or as an "insulator that does not allow charge movement. (Uses **P-1**, **P-3**, **P-4** and **P-5**.)

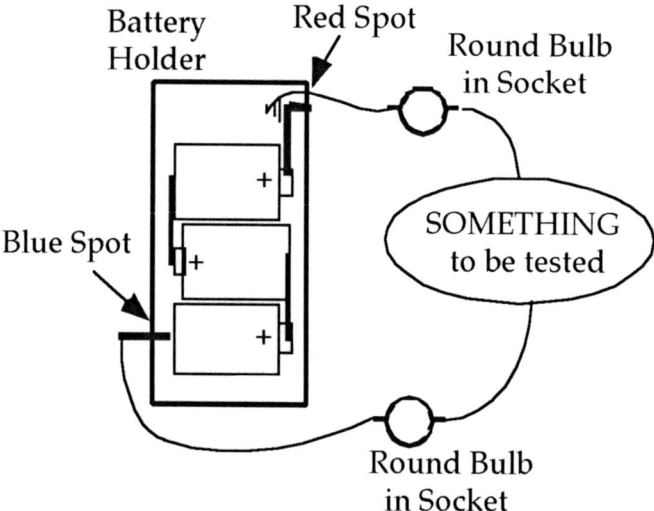

Fig. 5.3 Testing "something" as conductor or insulator

M-4: Resistance to charge flow

A resistor is added in a circuit with a single bulb and a 2-cell battery as shown in Fig. 5.4. Students explain the dimming of the bulb by thinking of a resistor as being "hard to get through". This flow-retarding characteristic property is called "resistance". Replacing the resistor by a second bulb again makes the original bulb dimmer – and thereby generates the idea that light bulbs also have resistance. The second bulb is introduced after students have observed the effect of a resistor on the brightness of a single bulb. If the second bulb is introduced before the resistor, students may interpret the dimming of the first bulb as evidence that the two bulbs are sharing something that's coming out of the battery. (Uses **P-1**, **P-3**, **P-4** and **P-5**.)

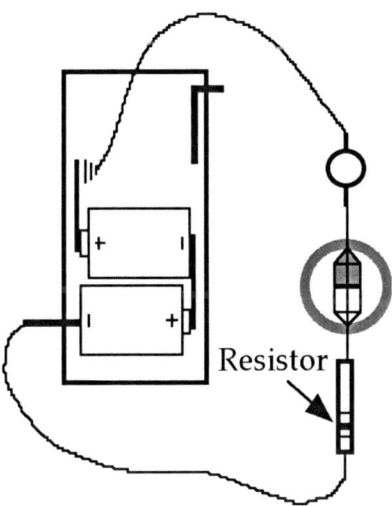

Fig. 5.4 Detecting effect of adding a resistor in a circuit

5.3.2 Moving Beyond the Battery Emission Model

Observing bulb lighting when a capacitor is placed between the pair of bulbs in Fig. 5.1 provides two major challenges to students' initial battery emission model of circuit operation. The following observations put students on course to *identifying a causal agent* in the wires:

- During capacitor charging, bulb lighting in a broken circuit provides evidence that the battery is not the only place where the charge that's moving in the wires originates.
- During capacitor discharging, bulb lighting without a battery in the circuit provides evidence that batteries are not the only causal agents that make charge move in wires.

Continued investigation of the influence of capacitors on bulb lighting helps students conceptualize what's moving as a compressible fluid that exists in all metal parts of a circuit – in capacitor plates and in wires as well as in batteries – and that the non-battery causal agent is a pressure-like condition in that fluid. Professionals call this condition "electric potential". The CASTLE curriculum calls it "electric pressure" – the term preferred by students, presumably because it supports transfer of kinesthetic imagery from experiences involving air pressure.

The key idea in this compressible fluid model is that PRESSURE in the conducting circuit components is higher/lower than normal in components where charge is compressed/depleted. A battery moves charge internally – out of the terminal labeled (–) and into the terminal labeled (+) – thereby compressing/depleting charge in the +/– terminal and creating above/below normal pressure in the +/– terminal. The pressure difference in the battery terminals drives charge flow from higher to lower pressure in the external circuit. The pressure difference in the two wires connected to a resistor drives charge flow through the resistor from higher to lower pressure.

M-5: Metal components also contain charge

The flow directions show by arrows in Fig. 5.5a are based on evidence from deflections of a magnetic compass placed under each wire. They show that the charge moving through the bottom bulb is coming *out of the metal* that the bottom capacitor plate is made of. Thus, the charge that moves in the wires originates not only in batteries but also *in the metal* that capacitor plates are made of – and, by inference, in the metal that the wires are made of.

Teachers draw Fig. 5.5b to foster simultaneous visualization of flow out of the bottom plate with depletion of its charge and flow into the top plate, which eventually "fills up" and explains why charging stops. The imagery of water flowing out of one tank and into another serves well for the moment. (Uses **P-1**, **P-2**, **P-3**, **P-4**, **P-5**, **P-6**, **P-7**.)

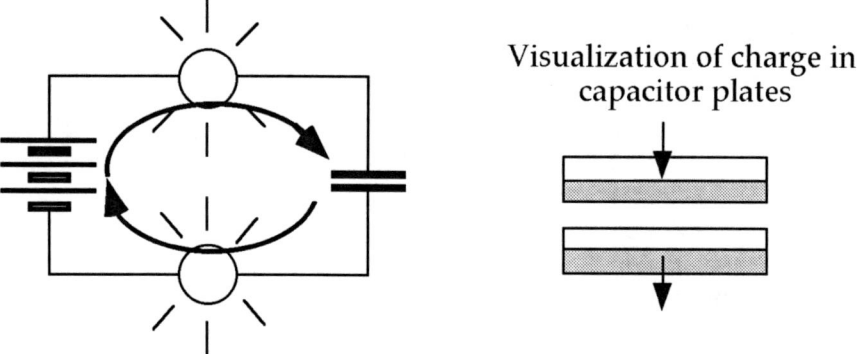

Fig. 5.5 (a) Charge flow through bulbs to and from a capacitor. **(b)** Visualizing flow into and out of capacitor plates

M-6: Charge distinguished from energy

The belief that glowing bulbs are getting using up something that comes to them from the battery is a valid intuition about ENERGY. This validity is acknowledged by showing how energy transferred to bulbs differs from CHARGE moving through bulbs.

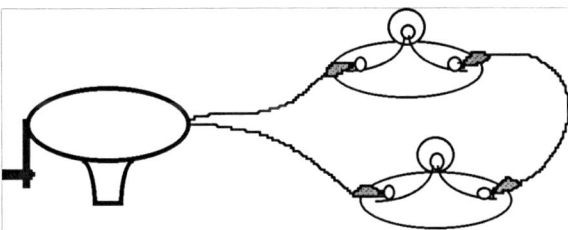

Fig. 5.6 Bulb lighting by a hand-cranked hand generator

Figure 5.6 shows a hand-cranked generator lighting the same bulbs that the battery formerly lit. In this experiment, the ENERGY that enables the bulbs to light comes from human muscle and gets *to* the bulbs by a non-circuit route. This makes it clear that energy stored in a battery is fundamentally different from CHARGE, which is a constituent of all the metal parts of a circuit and moves through bulb filaments *without being used up*. (Uses **P-1**, **P-3**, **P-4**, **P-5**, **P-6**, **P-7**.)

M-7: Showing that charge is compressible

Why does capacitor charging through the light bulbs shown in Fig. 5.7a eventually stop, as shown in Fig. 5.7b? Is it because the bottom plate becomes empty and can't give out any more? Is it because the top plate becomes full and can't take in any more? Adding the extra battery pack and asking students to predict what will happen as shown in Fig. 5.7c causes surprise when the bulbs light up again – clear evidence that the answer to both questions is NO. When both battery packs are removed and the capacitor is discharged, extra-bright bulbs reveal an extra-large flow rate driven by an extra-strong causal agent pushing charge back out of the top capacitor plate. This suggests that charging is stopped by a buildup of *pressure* – caused by *compression* due to inflow in the top plate -- that opposes charging more and more strongly. When the capacitor is subsequently discharged, charge flow is driven by this pressure. The pressure difference produced by charge compression/depletion is now the causal agent of current propulsion. (Uses **P-2**, **P-3**, **P-4**, **P-5**, **P-6**, **P-7**.)

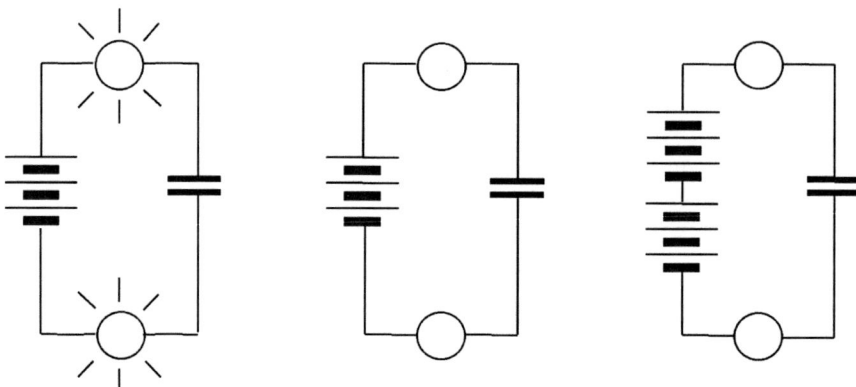

Fig. 5.7 (a) Capacitor is charging; **(b)** Charging is completed; **(c)** Another battery is added

M-8: Discrepant question

What pushes charge out of the bottom capacitor plate and into the bottom battery terminal as indicated in Fig. 5.5a. A simple answer is now available:

- The initially *normal* pressure in the bottom capacitor plate can drive charge out, and into the battery, only if there is *below*-normal pressure in the bottom battery terminal.
- The battery must create below-normal pressure in its bottom terminal by moving charge out, and into the top terminal, which creates *above*-normal pressure in the top terminal.

But students often lack an intuition that "normal" air pushes – and so lack imagery that could enable them to simulate normal pressure pushing charge out of the bottom capacitor plate. This situation is easily remedied using an "air capacitor", made of two plastic jars separated by an elastic membrane (part of a balloon), as shown in Fig. 5.8a. Figure 5.8b shows an air capacitor after a student has inhaled air from the right side – when normal air pressure in the left side can be observed pushing the membrane to the right. (Uses **P-2**, **P-3**, **P-4**, **P-5**, **P-6**, **P-7**, **P-8**.)

Fig. 5.8a Air capacitor with normal pressure on both sides

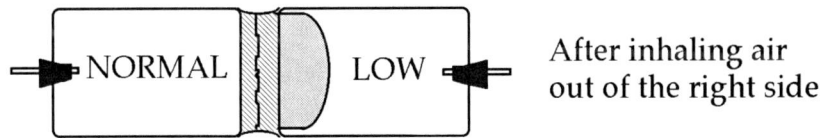

Fig. 5.8b Normal pressure pushing air toward below normal

A circuit is now regarded as having pressure in all metal components. Fig. 5.8c predicts

- a pressure difference pushing charge high-to-normal through the top bulb
- a pressure difference pushing charge normal-to-low through the bottom bulb

Thus the compressible fluid model, together with the battery modeled as a device that maintains a pressure difference in its terminals, predicts the observed clockwise flow pattern shown in Fig. 5.5a – out of the top battery terminal and into the bottom terminal.

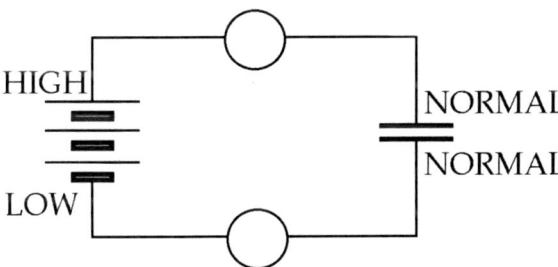

Fig. 5.8c Pressure values when capacitor charging begins

M-9: Pressure values in wires

After observing equal bulb brightness for all four bulbs in Fig. 5.9a, students are asked to use the color code at the right to assign pressure values in the battery terminals and wires. They easily conclude that the only realistic pressure values are those shown in Fig. 5.9a.

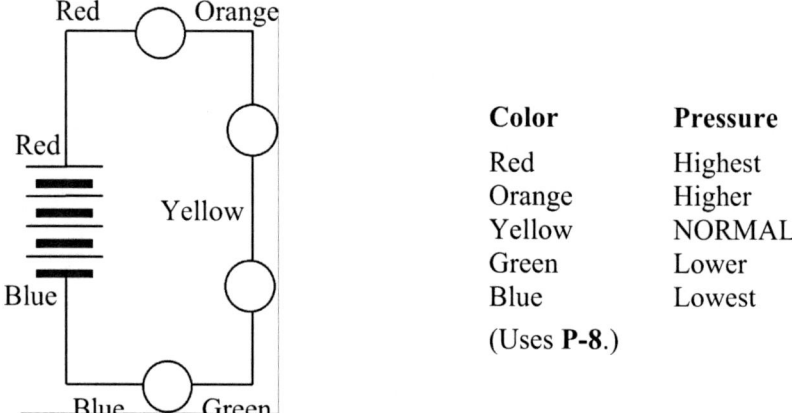

Color	Pressure
Red	Highest
Orange	Higher
Yellow	NORMAL
Green	Lower
Blue	Lowest

(Uses **P-8**.)

Fig. 5.9a Charging of wires begins when the circuit is closed

But how do different wires acquire different pressure values? The dynamic imagery associated with Fig. 5.5a, which explains the process of capacitor charging, can be used to run a mental model of "wire charging" in which circuit wires are containers in which compression and depletion occurs. This process is tracked by image enhancement in Figs. 5.9b and 5.9c, in which arrowtails represent coarse-grained quantitative

- charge flow through the top and bottom bulbs declining over time
- charge flow through the two middle bulbs growing over time
- equal charge flow through all bulbs as the final steady-state condition

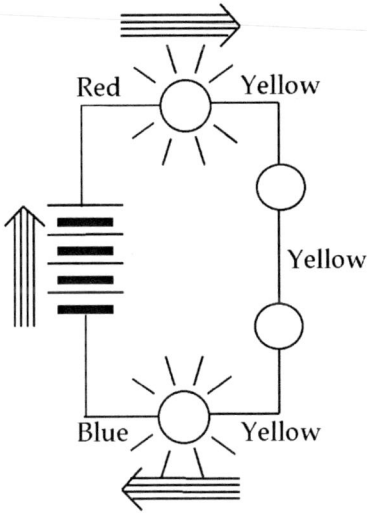

Fig. 5.9b Flow rates in wires have become more nearly equal

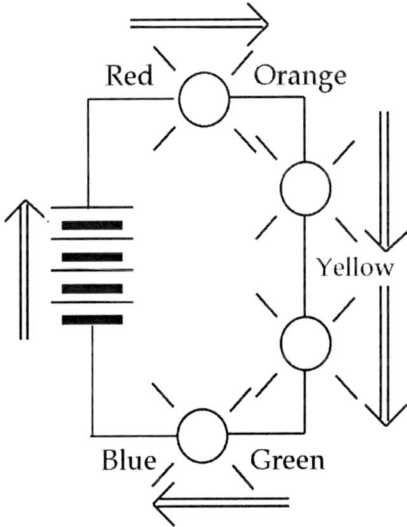

Fig. 5.9c Charging of wires ends when the flow rates equalize

The real-world transient "wire charging" process represented in Fig. 5.9b and 5.9c is much too rapid for human perception to observe a sequence of stages of bulb lighting. But stages of wire charging and discharging may be envisioned while using dynamic imagery to run the model mentally. And the conclusions can be supported by imagery enhancements on circuit diagrams that represent bulb brightness with starbursts, charge flow rate with arrowtails, and pressure values with color coding as in Figs. 5.9b and 5.9c.

M-10: Series voltage division

If all four bulbs in Fig. 5.9a have the same resistance, they will all glow with the same final brightness as in Fig. 5.9c – and the pressure color code will stand for the same pressure difference across each bulb. But what happens if, say, the two middle bulbs have greater resistance than the top and bottom bulbs?

Students can predict the course of the transient "wire charging" process by running their dynamical model. They understand that this is a transient process that ends when there is equal charge flow through all four bulbs. The prediction is that the process ends with a greater pressure difference across the bulbs with higher resistance than across those with lower resistance – in order to drive the same steady-state charge flow through all four bulbs. (The color code can be recalibrated to allow for unequal pressure differences.)

The conclusion that there must be a greater pressure difference across the bulbs with greater resistance provides a vivid inference of *series voltage division* – a principle that is widely used in circuit analysis and trouble shooting. It demonstrates the predictive ability of a visualizable model of pressure-driven current propulsion in circuits, supported by class discussion using appropriate imagery enhancement. (Uses **P-5, P-6, P-8**.)

5.4 Core Conceptual Goal of the Curriculum

Through the above experiences, the CASTLE curriculum aspires to enable teen-age students to build and reason with a visualizable model of current propulsion in circuits, based on knowing where mobile charge is located and the higher/lower pressure values associated with degrees of compression/depletion:

- Circuit wires are made of metal and contain mobile charge, like capacitor plates. They are not hollow pipes through which charge sent out by a battery is moving.
- A battery moves charge internally, creating high/low pressure in its +/– terminals.

- Charge flow driven by this pressure difference causes compression/depletion and creates HIGH/LOW pressure in the mobile charge in circuit wires connected to it.
- Mobile charge in a bulb filament is pushed from higher to lower pressure by the pressure difference in the wires that are connected to opposite sides of the bulb.

5.5 Unique Features of the Curriculum

I suggest that lack of attention by conventional electricity curricula to the initial student battery emission model, and to qualitative characteristics of electric circuits that could challenge this model, is an important contributor to student difficulty reasoning about circuits. To remedy this situation, the CASTLE curriculum incorporates features designed to make students aware of the gaps in their understanding and stimulate them to invent the means for filling these gaps.

5.5.1 Using Capacitors in Bulb-lighting Experiments

Adding a capacitor in series with light bulbs and a battery allows students to observe bulb lighting in a *broken* circuit while the capacitor is charging. This is evidence that the CHARGE moving in the circuit wires

- flows out of one capacitor plate, and is thus a *constituent* of the metal the plate is made of
- flows into the other capacitor plate, where its infusion causes crowding and *compression*

Discharging the capacitor through the bulbs allows students to observe bulb lighting with *no battery* to make charge move. This is evidence that the causal agent of charge flow resides not only in batteries – as beginners believe – but also in something that builds up in wires and other conducting circuit components. Dynamic imagery of the charging process suggests that this causal agent is PRESSURE in the capacitor plates, which

- pushes charge out of the plate with HIGH pressure caused by charge compression
- pushes charge into the plate with LOW pressure resulting from charge depletion

5.5.2 Using Air Compression and Pressure Analogies

Air is the compressible fluid that students are most likely to have had concrete experiences with. It is also the compressible fluid for which additional experiences are most readily available that can enable students to experience compression and its consequences concretely – to feel the increase of pressure produced by compression and to observe the movement caused by pressure. The means for doing this include balloons, air syringes and a teacher-made "air capacitor" – two membrane-separated chambers containing air with tubes for inflow and outflow of air. Dynamic imagery generated by these experiences appears well suited for transfer into circuit contexts.

5.5.3 Mobile Charge Being Always Present in Wires

In expert terms, electric potential is a condition in a distribution of electric charge in the circuit – and thus a condition in the mobile charge that inhabits circuit wires. A basic conception in circuit theory is that charge flow through a resistor is driven by the difference of electric potential in a pair of wires connected to the resistor.

In the CASTLE curriculum, electric potential is visualized as *pressure in a compressible fluid* of mobile charge that is *always present in wires*. Students know mobile charge is always present in the circuit wires because (a) they have empirical evidence that it is always present in capacitor plates and (b) they know that wires are made of the same metals as capacitor plates.

Students' initial battery emission model may include circuit wires being *hollow*, and empty except when a battery is sending something out through them to maintain bulb lighting. But it is not possible to assign electric potential values to wires that do not contain any mobile charge for the electric potential to be a condition in. Thinking of wires as being empty when there is no charge flow can lead to absurd conclusions – e.g. that there is zero potential difference across an open switch. Mobile charge being *always present in wires* is thus key to effective reasoning with electric potential in circuits.

5.5.4 Unpacking Hidden Issues

The mathematical modeling that guides conventional electricity curricula leaves important issues of qualitative understanding hidden – unarticulated and beyond the reach of empirical investigation. An important example is learning how different wires connecting resistors in series acquire the different pressure values needed to drive the same charge flow rate through each resistor. As shown above using bulbs as resistors, these pressure values are acquired during a transient process involving compression and depletion of charge in the wires that occurs after a battery is connected. The pressure changes in this process occur too quickly to be observable in practical circuits – and they are ignored in conventional curricula. The CASTLE curriculum helps students generate imagery for simulating this process by connecting a capacitor in parallel with series bulbs to make human perception possible by lengthening the transient bulb lighting time.

5.5.5 Separating Charge and Energy Concepts

The observation of bulb lighting while a capacitor is charging and discharging initiates a revolution in students' conceptions of what moves in wires and what makes it move. These experiences powerfully challenge their battery emission model, which is inappropriate for the CHARGE that moves in the wires since it is a property of the metal that the wires are made of rather than something sent out from a battery.

It is nevertheless important to acknowledge the validity of students' deeply held conviction that *something* in the battery is making a one-way trip to glowing bulbs and is being used up in the bulbs. What's needed is evidence that the ENERGY being *used up in* the bulbs is different from the mobile charge that is *passing through* the bulbs and is being recycled around the circuit. This is done at two levels.

1. Experiences with a Genecon generate imagery of energy transfer in *non*-circuit paths.
2. Experiences with a hula-hoop generate imagery for a model with charge *and* energy.

The conception of energy that emerges is an enabling agent that is *transferred to* lit bulbs by the action of pressure differences, which are the casual agents that make charge *pass through* the bulbs.

5.6 Why Begin Electricity with Circuits?

Conventional electricity courses in high schools and colleges begin with electrostatics and introduce a distant-action force law based on vector electric field as causal agent. They define electric potential mathematically in terms of electric field – and then invite students to transfer their knowledge of electrostatics into circuit contexts and use this definition of potential difference to reason about current propulsion in circuits. Research has consistently shown that students taught this way have difficulty predicting and explaining qualitative circuit behaviors. (Fredette & Lochhead, 1980; Cohen, Eylon, & Ganiel, 1983; Duit et al., 1985; Shipstone et al., 1988; Engelhardt & Beichner, 2004). The salient findings are that students confuse current and potential difference, and use sequential reasoning as a substitute for reasoning with potential difference as current driving agent (Closset, 1983).

Professionals regard a robust conception of electric potential as key to effective reasoning about circuits. Why is conventional electricity instruction failing to achieve this goal? My view is that this mode of instruction creates an isolated domain of mathematical relationships in which the causal agent of current propulsion (potential difference) is defined with two major deficits:

1. It lacks a strongly intuited model for causal agency in circuits. It ignores the air pressure analogy, which played an important role historically. (Steinberg, to appear).
2. It lacks the efficient explanatory power that comes from having a runnable model. The idea of electric pressure differences driving charge flow is imageable and runnable, and provides a basis for making rapid chains of inferences. (Clement & Steinberg, 2002).

In contrast, four decades of classroom experience with batteries-and-bulbs pedagogy (Elementary Science Study, 1962) in middle school, high schools and colleges has shown that:

- Beginners of all ages are intensely interested in experimenting with bulb lighting.
- Ninth graders already have the intuition that "something is moving in the wires".
- Students have useful intuitions about air pressure and flow, which with some additional development can become the starting point for constructing runnable imagistic models of charge flow in circuits driven by pressure differences in the metallic circuit components.

The CASTLE curriculum builds on this more promising foundation. The runnable models that it helps students access can provide flexibility for managing transfer to unfamiliar problems, using imagistic simulation and spatial reasoning systems that have more built – in flexibility than formal algorithms. (Clement, 2003).

5.7 Effectiveness of the *Castle* Curriculum

The following data from 14 American high schools are from a test on ability to reason about simple circuits, which compared student performance in classes using conventional curricula with classes using the CASTLE curriculum. (Brown, 1994) There were no capacitors in the test circuits, in order not to disadvantage students in the classes with conventional instruction. The questions were designed to be somewhat difficult, in order to avoid a ceiling effect. Teachers in all classes had a reputation for excellence.

The numbers in the following table show percentages of correct answers to questions involving circuits of batteries, light bulbs and switches. Students selected a confidence level for each question, ranging from 1 (just a guess) to 5 (I'm sure I'm right), and the responses were converted to a percentage of 5. The data provide evidence that the new model based curriculum is superior to conventional electricity curricula in ability to foster both conceptual gains and confidence gains.

% correct answers (p<,001)	Pre	Post	Gain
Conventional curriculum	32	36	4
CASTLE curriculum	30	50	20

Confidence levels (p<,001)	Pre	Post	Gain
Conventional curriculum	56	63	7
CASTLE curriculum	53	74	21

Assuming correct answers to be a measure of conceptual gain, these data show that classes using the CASTLE curriculum make made conceptual gains 5 times as large as those using a conventional curriculum. Confidence gains for classes using the CASTLE curriculum were 3 times as large as for those using a conventional curriculum. The confidence gain differential was especially large for the female students (25 CASTLE to 3 conventional).

5.8 Conclusion

An attempt was made at the beginning of this paper to summarize the approach used in a set of principles for curriculum design. My impression is that these principles are rarely applied systematically to curriculum design.

The application of Incremental Modification (**P-4**) produced a large number of steps that are not found in conventional electricity curricula. The development team identified these steps and their Anchoring Conception (**P-1**) by talking to students, holding tutoring trials, and teaching trial versions of the curriculum. These efforts identified feasible target models and sub-problems (not all of them anticipated ahead of time) encountered in building these models. We found that teachers who are serious about fostering solid conceptual understanding by students typically agree that a coordinated sequence of many small steps is required. We call this sequence a Critical Learning Pathway. An important feature of the pathway is that a model is developed in small pieces, so that misconceptions and other issues can be dealt with incrementally.

Of special interest to researchers are the principles focusing on imagery: Representation in Imagery (**P-5**); Transfer of Dynamic Imagery (**P-6**); and Imagery Enhancement (**P-8**). These are perhaps the least recognized principles in science education at this time. Theoretically, using dynamic imagery to represent physical relationships has major advantages for beginners over the mathematical representation favored in conventional electricity courses:

- Multiple constraints in a physical system can be represented in the same image.
- Imagistic relationships allow analysis by means of concrete spatial reasoning.
- Imagistic simulations can evaluate model modifications and solve novel problems.

Finding, implementing, and studying other manifestations of these imagistic principles is an important task for future research and development.

References

Brown, D. E., (1994). A statistical study of CASTLE curriculum effectiveness in 14 high schools for U.S. National Science Foundation (analysis available by request).

Clement, J., (2008). *Creative model construction in scientists and students: Imagery, analogy, and mental simulation.* Dordrecht: Springer.

Clement, J., (2005). Thought experiments and imagery in expert protocols, *in this volume.*

Clement, J., (2003). Imagistic simulation in scientific model construction, In Alterman, R. & Kirsh, D. (Eds.), *Proceedings of the Twenty-fifth Annual Conference of the Cognitive Science Society*, 25, 258–263, Erlbaum, Mahwah NJ.

Clement, J., (1994). Use of physical intuition and imagistic simulation in expert problem solving, in Tirosh, D (Ed.), *Implicit and explicit knowledge*, Ablex Publishing Corp., Norwood NJ.

Clement, J. and Steinberg, M. S., (2002). Step-wise evolution of models of electric circuits: a "learning-aloud" case study, *Journal of the Learning Sciences, 11 (4)*, 380–452.

Closset, J.-L., (1983). *Le raisonnement sequentiel en electrocinetique*, doctoral thesis, University of Paris VII.

Cohen, R., Eylon, B., & Ganiel, U., (1983). Potential difference in simple electric circuits: a study of students' concepts, *American Journal of Physics*, 51, 407–412.

Duit, R. et al. (Eds.), (1985), *Aspects of understanding electricity*, Schmidt & Klaunig, Kiel.

Elementary science study, (1962). *Batteries, And bulbs,* New York: McGraw-Hill.

Engelhardt, P. V., & Beichner, R. J., (2004). Students' understanding of direct current resistive electrical circuits, *American Journal of Physics, 72*, 98–115.

Fredette, N., & Lochhead, J., 1980, Student conceptions of simple circuits, *Physics Teacher, 18*, 303–316.

Niedderer, H. and Goldberg, F, (1996). *Learning processes in electrical circuits.* Paper presented at Annual Meeting of NARST.

Shipstone, D. et al., (1988). A study of students' understanding of electricity in five European countries, *International Journal of Science Education*, 10, 303–316.

Steinberg, M. S. & Wainwright, C. L., (1993). Using models to teach electricity – the CASTLE project, *Physics Teacher, 31*, 353–357.

Steinberg, M. S. et al., (2007). *Electricity visualized*, PASCO Scientific, Roseville CA. (Download from www.pasco.com after calling 800-772-8700 for access information.).

Steinberg, M. S. (to appear). *Inventing electric potential, special issue of Foundations of Science*, L. Magnani ed.

Chapter 6
Case Study of Model Evolution in Electricity: Learning from Both Observations and Analogies

John Clement

University of Massachusetts, Amherst

Melvin S. Steinberg

Smith College

6.1 Introduction

This chapter focuses on data from a tutoring case study in high school level instruction on electric circuits. A model evolution approach to instruction is described that works within a cycle of model generation, evaluation, and revision. Case studies of transcripts from the lessons allow one to develop diagrammatic representations of the learning and teaching processes involved.

The data base for this study is a set of tutoring interviews with a student, who we shall call Susan, who was 16 years old and who had completed her junior year in high school. Her teachers characterized her as having above average but not highest level ability in science. Susan had taken a course in chemistry but had not yet taken a course in physics. The instructional techniques included the use of both analogies and observations of real circuits constructed by the student and the teacher. The diagrams map the interplay between these instructional modes, and their effects on the student's evolving model. They also illustrate important differences between source analogies and target models. Diagrams of learning processes at this scale are rarely mapped out carefully on the basis of learning theory. If such general mapping tools can be developed it should help us develop more detailed models of conceptual change. Additionally, having visual and verbal languages for these planning domains could be of considerable value for curriculum development. One advantage of conducting a case study of a one-on-one tutoring interview, rather than of a full classroom, is that much clearer data on learning processes is obtainable in a systematic way from each student as they "learn aloud" (Clement, 2002).

6.2 Learning in the Domain of Electric Circuits

In Susan's first session she was asked to think aloud as she completed a pretest on electric circuits. She did this again with an identical posttest in her last session. The five intervening tutoring sessions were spread over a period of two weeks and lasted approximately eight hours in total. Susan was also assigned a homework problem after most of these sessions. During the tutoring Susan was asked to think aloud as she set up and observed experiments with circuits, explained events, solved problems, reacted to the tutor's comments, and completed color coding for "electric pressure" (potential) values in circuit diagrams.

Pretest. Susan's pretest revealed that she had very little academic knowledge of electric circuits. In simple circuits with a battery and one or two bulbs, she felt that electricity must flow somehow from a battery to a bulb, but was unsure about the path it would take. On more difficult questions, she said she had no idea about what would happen, and the instructor reassured her that this was all right and not unusual.

Background to the Protocol. In this chapter we will describe episodes initiated by transient bulb lighting events to which Susan, reacted with strong expressions of surprise. The transcripts presented from this five day intervention are necessarily only a small piece of the entire intervention. Prior to these episodes, Susan had learned in hands-on experiences: (a) how to hook up simple circuits of flashlight batteries and light bulbs, and (b) how to find out whether materials used in circuits are conductors or insulators. When asked what she thought might be happening in the wires during bulb lighting, she began talking about *something moving* through the wires from the battery to the lighted bulbs. At first Susan talked about "positive and negative currents" moving out of the battery terminals labeled "+" and "−". When she seemed blocked by a morass of questions associated with currents flowing in opposite directions to two bulbs connected in series, the tutor suggested trying the simpler idea that bulb lighting is associated with something moving in a *single* direction in each wire. This was confirmed when Susan used a compass placed under each wire to determine (relative) directions of movement in the wires. The instructor recommended using the name "charge" for "whatever-it-is that's moving" in the wires.

Surprises. The circuit that generated Susan's first surprise is shown in Fig. 6.1. All earlier investigations had used the two-bulb circuit shown in Fig. 6.1 *without the capacitor*. Before introducing the circuit in Fig. 6.1, the tutor described the layered conductor-insulator-conductor architecture of a capacitor: two conducting sheets of metal (called "plates") are separated by an insulating sheet sandwiched between them so that there is *no direct contact* between the plates.

6.3 First Model Building Cycle

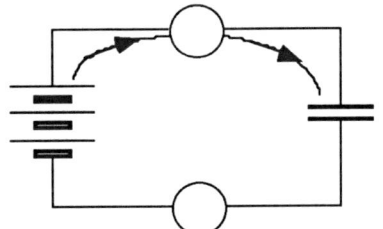

INITIAL CONCEPT: Battery is the only part of circuit where what's moving originates.

Fig. 6.1 Capacitor charging – Bulbs light briefly for 1–2 seconds and then go out in a circuit with a capacitor connected to two bulbs and a battery

6.3.1 Modeling Cycle 1

Susan, therefore, expected that the bulbs would not light. However, when connected, they did light for 1–2 seconds and then went out. In responding to probes from the tutor, Susan decided that this meant current must flow into one side of the capacitor and stay there, like a fluid flowing into a container. Because the bottom bulb lit too, Susan agreed that there must have been current coming out of the other side of the capacitor as well.

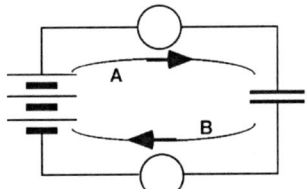 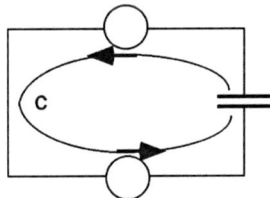

Fig. 6.2 (a) Charging capacitor; **(b)** Discharging capacitor. Charge *moves back out* of the top plate after the battery is removed and wires are reconnected

6.3.2 Modeling Cycle 2

Removing the battery and connecting the wires that were attached to it initiates a new round of bulb lighting for 1–2 seconds. Susan used a compass to find the direction of movement in each wire during capacitor discharging. She was surprised to discover that charge that had moved along path A in Fig. 6.2a from the battery into the top capacitor plate during charging is now *coming back out*, moving along path C in Fig. 6.2b, as follows:

(2.1) SUSAN "Once you take it [the circuit] apart, you're eliminating the battery pack. And I was thinking why the current is gonna move in the other direction."

(2.2) "...there's no place where the charge originates...to get it to go in the other direction."

Here we have evidence for Susan sensing another gap in her model – where in line 2.1 she asks what could be the causal reason for charge to move back out of the top capacitor plate. In line 2.2 she seems troubled by the absence of any agent that can make the charge move.

The tutor exploited Susan's emerging need for a causal agent to introduce a potentially useful analogy:

(2.3) TUTOR "Now I'm going to talk about something completely different, which is going to seem to be unconnected..."

(2.4) "I would like you to think about an automobile tire..."

(2.5) "What happens if we put a nail into it...?"

(2.6) SUSAN "Then you're going to allow an escape for the air..."

(2.7) TUTOR "Why does the air escape?..."

(2.8) SUSAN "Because you've got the great pressurized air inside of your tire..."

(2.9) "The air is gonna want to move to an area where there's less pressure..."

She then begins applying these ideas about a tire containing air to a capacitor plate containing charge:

(2.14) SUSAN "Uh, uh. I was just trying to hang onto everything."

(2.15) TUTOR "Tell me what you're thinking."

(2.16) SUSAN "You're never going to be completely empty of the charge. You're always going to have some charge..."
"Whatever metal, or whatever you have, there's always gonna be some amount there..."

(2.17) TUTOR "You're on a good track..."

(2.20) TUTOR "Why does this charge move back through these bulbs during discharging? What provokes it?"

(2.21) SUSAN "Once you take the battery cell out...you don't necessarily have that pump forcing the air in..."

(2.22) "And that's sort of like...punching a hole in a tire, or whatever, and letting it go back in the other direction."

(2.23) TUTOR "What part is like punching a hole in the tire?"

(2.24) SUSAN "Kind of discharging. Connecting two wires so that..."

(2.25) TUTOR "Connecting the two wires is like punching a hole in the tire?"

(2.26) SUSAN "Yeah..."

It appears from the transcript that Susan is now moving toward a conception of charge in a capacitor plate as being like air under pressure. That is, she was able to map and transfer certain elements of the tire case into the circuit case. The discrepant event of the bulb lighting without a battery seems to have motivated this conceptual change. (Notice that the tire itself is not incorporated into the explanatory model for the circuit. Rather, it is only certain elements of the analog conception involving air pressure that are incorporated into a revised model of the circuit. Thus, we make a distinction in this instance between source analogies and explanatory models.)

6.3.3 Modeling Cycle 3

In a third modeling cycle, Susan became puzzled about why charge left the bottom plate of the capacitor during charging back in the original experiment in Fig. 6.1. Why should charge be forced from the lower plate of the capacitor toward the battery? The tutor engaged Susan in discussion about the analogy, asking what will happen if one pumps the air out of a jelly jar and then punctures the top of the jar. This enabled Susan to imagine outside air at normal pressure pushing into the region of below-normal pressure inside the jar. To regard this as an analogy for what is happening in path B, Susan imagined a region of below-normal pressure in the bottom terminal of the battery. The tutor explained that to make that happen, the battery pumps charge out of the bottom terminal – and into the top terminal, where it would produce *below*-normal and *above*-normal pressures. Accepting this possibility meant that Susan abandoned the original conception of a battery as a squirt-can in order to view it as more like a pump with an intake and an output.

6.3.4 Evidence for Conceptual Understanding

Susan was given a post test question that is a genuine transfer problem in the sense that it involves new features she had not dealt with earlier during the instruction. The problem involves 4 resistors, and has a central path not in a perimeter of the circuit. As a result, it is a fairly stringent test of the depth of her understanding. Her transcript from this problem indicates that she has an unusually robust conception of electric potential, which she can apply to unfamiliar situations of fairly high complexity (see Clement & Steinberg, 2002 for a more detailed discussion of this transcript.) She is able to reason using pressure differences in wires as the current driving agent in the circuit. Other studies have found that potential difference is a stubbornly difficult concept that typically remains unlearned after instructional interventions (Niedderer & Goldberg, 1996) (Cohen, Eylon, & Ganiel, 1983). She gives a coherent explanation in her own words of the pressure changes that will occur as the original system passes through a transient process to a state of steady flow. This suggests that Susan has gone through a considerable conceptual change compared to her early circuit models and has constructed a conceptual understanding of several aspects of electric circuit mechanisms.

6.3.5 Diagramming Susan's Learning Cycles

We propose that Susan's learning consisted centrally of several cycles of model construction. Figure 6.3 shows the hypothesized form of these cycles as they affect conceptions in the mind of the student. The three rows in the figure, from top to bottom, represent the student's Prior Knowledge (tapped by analogies), Evolving Explanatory Model, and Observations. The middle row shows the development of the student's model with time going from left to right and begins with the student's own initial model labeled M1, used to form the prediction that neither bulb would light for the first example of a circuit presented in Fig. 6.1. The prediction conflicted with the student's Observation 1, shown in the bottom row. The resulting dissonance is symbolized by the zig-zag line between Model 1 and Observation 1. This dissonance motivates the student's construction of Model 2, as a modification of Model 1.

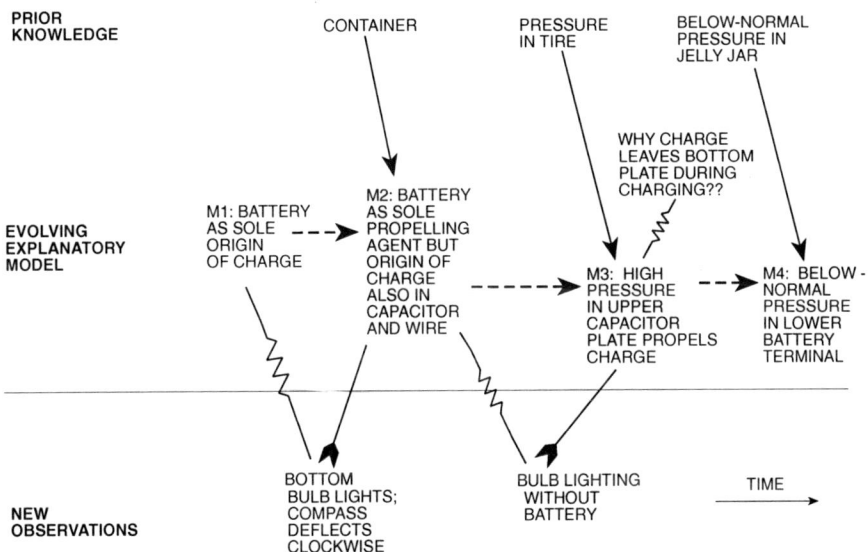

Fig. 6.3 Evolution of Susan's mental model for circuits as influenced by observations and analogies

Useful student preconceptions that are activated by the presented analogous cases are shown in the top row labeled Prior Knowledge. The modification of Model 1 to generate Model 2 is successful if it is constructed so that it explains Observation 1 more adequately than Model 1 did. Model 2 is an example of what we call an "Intermediate Model" that is only partially correct. Nevertheless it represents progress over Model 1. In theory the cycle can be repeated as many times as needed. (Students will also use other elements of prior knowledge about bulbs, wires, batteries, and air that are not included in the analog conceptions, but these are not depicted.) In this diagram, the entries are inferred from events in the transcript, and the center row represents our cognitive model of the student's evolving model, based on student drawings and statements. (Many more cycles are required in the full CASTLE curriculum.) We refer to the middle row of Fig. 6.3 as showing an abbreviated summary of Susan's *learning pathway* from her initial model, through a series of intermediate models, to a target model.

We view the lesson sequence as exhibiting an overall strategy which we call learning via model evolution. It involves the coordinated use of multiple analogies, discrepant events, student explanations and discussions, and model evaluation and revision, in an attempt to lead the student through a series of more and more adequate models. Although Fig. 6.3 suggests the order in which teaching "moves" were implemented, it goes beyond this in representing a theory of the cognitive events taking place in Susan in response to, and sometimes in spite of, the tutor. In this case Susan, not the tutor, raised the questions of why charge leaves the bottom plate during charging. This led to a spontaneous analogy constructed by the teacher, which later became incorporated into the published curriculum.

6.3.5.1 Beyond Simpler Views of Conceptual Change

Figure 6.4b isolates the second cycle from Fig. 6.3 and compares it with a more simplistic view of conceptual change in Fig. 6.4a that has appeared previously in the conceptual change literature. In Fig. 6.4a a discrepant event conflicts with the student's initial Model 1, motivating it being discarded. An analogy then initiates a correct Model 2, which replaces Model 1. The distinguishing features of the more complex view of conceptual change in Fig. 6.4b are:

1. The change from a "remove and replace" (exchange) view of the change in the model to a "transformation" view that modifies small elements of the initial model to form an intermediate model M2. This is shown by the "modification" arrow in Fig. 6.4b.
2. The discrepant event *constrains the construction of the new model as well as creating dissonance with the old model*. This is shown by lines pointing from the discrepant event to the modified model in addition to the jagged lines showing dissonance with the previous model. The discrepant event plays a dual role; its function is not just to remove a prior misconception.

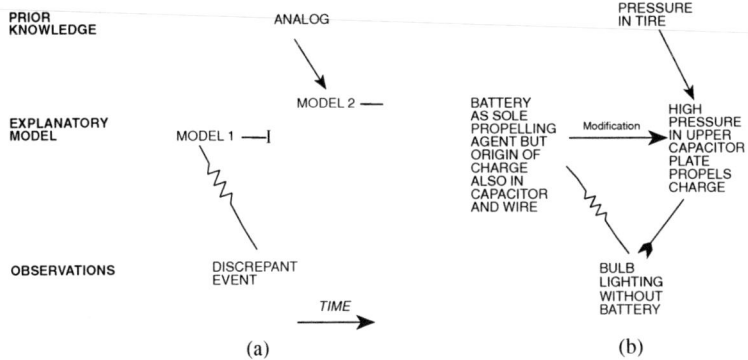

Fig 6.4 (a) Traditional view of conceptual change; (b) Revised view of conceptual change

Thus, in this case study we have seen how strategies such as analogies and discrepant events were able to produce conceptual changes in Susan's model of electric circuits. Although these are sometimes seen as isolated strategies that can cause conceptual change on their own, here we see them (1) being coordinated and (2) being part of a cyclical process of model evolution.

6.4 Learning Processes in the Case Study that Suggest General Instructional Principles

6.4.1 Overview

This section proposes general processes that we believe played a role in the learning. These include the evolutionary sequence of models and their revisions, the role of prior knowledge, specialized use of analogies, and the use of discrepant events.

6.4.2 An Evolutionary Sequence of Models and Revisions

The sophistication of Susan's explanations grew steadily during the instructional treatments, as shown in Fig. 6.3, to produce a sequence of progressively more expert-like models. This suggests a view of her learning process that has *model evolution* as its central feature, where students are able to build on knowledge that they had developed in earlier sections. Evolution occurs via generation, evaluation, and modification cycles or GEM cycles (Clement, 1989, 1993, 2008). Three cycles of evaluation and modification are shown in Fig. 6.3.

6.4.3 Learning is an Interaction Between Prior Knowledge and New Observations or Ideas

Observations alone were not viewed as sufficient for producing normative conceptual change. Positive sources of knowledge in the student are also used. The process relied on using a balance of rational-analogical (prior knowledge) as well as empirical sources of ideas rather than relying primarily one or the other. The three levels shown in Fig. 6.3 reflect the theoretical position that contributions are made both "from above" and "from below" as observations interact with prior conceptions. Thus the process fits an interactionist view of learning as empirically constrained, creative model construction rather than an empiricist view of learning as generalizing "upward" from observations.

6.4.4 Specialized Use of Analogies

Several features of analogy use in this tutoring session are notable in contrast to traditional ways of using analogy. In contrast to using single analogies only, *multiple analogies* are used to contribute elements to the evolving model. (See also Glynn, Doster, Nichols, & Hawkins, 1991; Spiro, et al., 1989.) In this view we *distinguish between explanatory models and source analogies*. In the case of the tire analogy in electric circuits, we saw that the tire itself was not incorporated into the model. Instead, one attempts to draw certain features selectively from each analogy in order to add a component to the evolving model. Although some researchers have tended to treat a model and a source analog as equivalent, we believe that researchers, teachers, and students need to distinguish an explanatory model from the particular analogous cases that contribute elements to its construction. This is seen most clearly in a case like this one where multiple analogies are used. This is not to say that single analogies are never useful (they were used on occasion for example in Camp and Clement, et al. (1994)). Rather it reflects our view that a single analogy would be insufficient for developing a model of this complexity.

6.4.5 Use of Discrepant Events

The first sources of dissatisfaction and model evaluation in the instructional sequence were the discrepant events used to motivate model revisions. We have seen that these did indeed violate the student's expectations, caused surprises, and were eventually followed by model revisions, meaning that the student was eventually able to explain the events satisfactorily. We modeled effects of the discrepant events as internal dissonance with an existing conception. These are shown as jagged lines in Fig. 6.3.

Another principle that can be formulated is that the discrepant events were designed to encourage the expansion of the student's model in small "mind sized" pieces. As opposed to trying to promote maximal dissonance, dissonance was promoted by the suggested experiments in *small, doable, and repairable steps*; that is, steps designed so that model modifications were not too large at any point. This also means that *timing* of such sources of model criticism is important. The sources were timed so that they were not introduced until the students had an initial partial model – i.e. they had the groundwork they needed to modify their model to remove the dissonance. Thus the material was carefully sequenced to find the next teachable structure element for the students' current state of model development.

The aim in this approach is for sources of dissonance to produce teachable moments rather than discouragement. This contrasts with a "remove and replace" strategy of first trying to eliminate the students' conceptions in the area completely by setting up a major conflict with the students' preconceptions immediately and then building up new knowledge from a blank slate. This changes the role of a discrepant event from that of a "conception remover" to a "conception improver"– a role of bringing one aspect of the student's conception or model into question, so that that aspect can then be changed to modify and improve the conception or model.

This decomposition of the teaching problem into smaller mind sized pieces leads to the use of *multiple discrepant events* and other sources of dissonance, as illustrated in Fig. 6.3. The sources of dissonance in the bottom row of Fig. 6.3 have lines pointing to the modified model as well as jagged lines connecting to the previous model. This signifies that they are not just in conflict with the previous model but are also constraints used in the construction of the modified model. Thus they make *a positive as well as a negative contribution*. (Steinberg & Clement, 1997).

In summary, model evaluation techniques are carefully structured and timed in this approach to produce an *optimal level of dissonance rather than a maximum level*. At the metacognitive level, we believe that the tutor discussed in this instance was successful in conveying the spirit of the following credo: An idea, and in particular an explanatory model, can be put under examination and tested and evaluated by students, rather than by an authority. It did this by conveying the importance of the general questions: Are you convinced? Does it make sense to YOU? Does it make enough sense to you so that you can explain it in your own words? The implicit metacognitive message is that students are able to, and have the right to, self-evaluate and modify their own models so that the material makes sense to them. Ways must also be found to make this happen in the classroom more reliably (see Hennessey, 1999; White & Frederiksen, 2000).

6.5 Conclusion

6.5.1 Model Evolution vs. Simpler Approaches

This sample lesson highlights the evolutionary approach that may be needed for building complex models. We believe that models of systems as complex as electric circuits cannot be constructed in a single intervention. The model evolves over a period of time through a longer chain of conceptual changes. The lesson diverges from other common approaches to conceptual change in several ways:

- Strategies aim to change an aspect of the student's model vs aiming to reject and replace the student's model
- Multiple discrepant events vs the use of one discrepant event
- Multiple analogies vs the use of a single analogy
- Use of rational sources of reasoning vs the use of experiment alone

6.5.2 Implications

We believe that in order to apply these ideas in the classroom, the cognitive strategies discussed in this chapter are essential but insufficient. They must be adapted and integrated with social and metacognitive strategies for learning in classrooms. In order to achieve this, however, ways must be found to encourage active learning on the part of each student. In classrooms that have adopted the electricity curriculum described in this paper, students are encouraged to construct explanations and arguments by working on experiments with each other in small groups organized by the teacher (Steinberg, et al., 2005). Between experiments, the teacher guides large-group Socratic discussions where meanings of terms are negotiated and alternative models are compared for explaining the lab observations. An attempt has been made to build learning pathways into the suggestions for discussion that reflect questions like the ones that occurred in this case study. Use of these strategies and others in full classrooms has been documented by Williams and Clement (2006).

The present approach illustrates the extensive preparation that can be involved in teaching for deep understanding in: (1) analyzing students' alternative conceptions developed both before and during instruction; and (2) finding and coordinating sequences of analogies and dissonance producing events that take these alternative conceptions into account. The approach appeared to create a viable learning pathway for gradual construction of a complex scientific model in the case studied. The most central strategies were the use of multiple analogies, multiple discrepant events, student explanations and discussions, and model evaluation and revision, in an attempt to help the student construct a series of more and more adequate models. However, the overall strategy was more complex than the use of individual methods. We view the strategies as fitting together as shown in Fig. 6.3, to form an overall strategy which we call learning via model evolution.

Our long term goals in this work are to frame grounded theoretical hypotheses about general teaching techniques, and for this, diagramming systems that can represent teaching strategies succinctly are an important tool. It is likely that having both visual and verbal languages for these planning domains will be valuable for curriculum developers and teachers as well as researchers.

References

Camp, C., Clement, J., Brown, D., Gonzalez, K., Kudukey, J. Minstrell, J., Schultz, K., Steinberg, M., Veneman, V., and Zietsman, A. (1994). *Preconceptions in mechanics: Lessons dealing with conceptual difficulties.* Dubuque, Iowa: Kendall Hunt.

Clement, J. (1989). Learning via model construction and criticism. In G. Glover, R. Ronning, & C. Reynolds (Eds.), *Handbook of creativity: Assessment, theory and research* (pp. 341–381). New York: Plenum.

Clement, J. (1993). Model construction and criticism cycles in expert reasoning. Proceedings of the Fifteenth Annual Meeting of the Cognitive Science Society, Hillsdale, NJ.

Clement, J. (2000). Analysis of clinical interviews: foundations and model viability. In A. E. Kelly, & R. Lesh, (Eds.), *Research methods in mathematics and science education* (pp. 341–385). Hillsdale, NJ: Erlbaum.

Clement, J., (2008) *Creative model construction in scientists and students: The role of imagery, analogy, and mental simulation.* Dordrect: Springer.

Clement, J. & Steinberg, M. S., 2002. Step-wise evolution of models of electric circuits: a "learning-aloud" case study, *Journal of Learning Science. 11* (4), 380–452.

Cohen, R., Eylon, B. S., & Ganiel, U. (1983). Potential difference and current in simple electric circuits: A study of students' concepts. *American Journal of Physics, 51,* 407–412.

Driver, R., H. Asoko, et al. (1994). "Constructing scientific knowledge in the classroom." *Educational Researcher 23*(7), 5–12.

Glynn, S., Doster, E., Nichols, K., & Hawkins, C. (April, 1991). *Teaching biology with analogies: Explaining key concepts.* Paper presented at the annual meeting of the American Educational Research Association, San Francisco.

Niedderer, H. & Goldberg, F. (1996). Learning processes in electric circuits. Paper presented at Annual meeting of the National Association for Research in Science Teaching.

Hennessey, G. (1999). Probing the dimensions of metacognition: Implications for conceptual change teaching-learning. *Proceedings of the National Association for Research in Science Teaching*, Boston.

Spiro, R. J., Feltovich, P. J., Coulson, R., & Anderson, D. K (1989). Multiple analogies for complex concepts: Antidotes for analogy-induced misconceptions in advanced knowledge acquisitions. In S. Vosniadou, & A. Ortony, (Eds.), *Similarity and analogical reasoning.* New York: Cambridge University Press.

Steinberg, M., & Clement, J. (1997). Constructive model evolution in the study of electric circuits. In R. Abrams (Ed.), *Proceedings, The Fourth International Seminar on Misconceptions Research.* Santa Cruz, CA: The Meaningful Learning Research Group.

Steinberg, M. S. et al. (2005). *Electricity Visualized — The CASTLE Project.* Roseville, CA: PASCO Scientific.

White, B. Y., & Frederiksen, J. R. (2000). Metacognitive facilitation: An approach to making scientific inquiry accessible to all. In J. Minstrell and E. van Zee (Eds.), *Inquiring into inquiry learning and teaching in science* (pp. 331–370). Washington, DC: American Association for the Advancement of Science.

Williams, G., & Clement, J. (2006). Strategy levels for guiding discussion to promote explanatory model construction in circuit electricity. AIP Conference Proceedings: 2006, Physics Education Research Conference, Syracuse University, Syracuse, New York, July 22–26.

Section III

Qualitative Research on Specific Strategies

Chapter 7
A Competition Strategy and Other Modes for Developing Mental Models in Large Group Discussion

Maria C. Núñez-Oviedo

Universidad de Conception

John Clement

University of Massachusetts

7.1 Introduction

This chapter examines three examples of large-group discussions with middle school students in the area of respiration. Several modes of teaching to foster model construction are identified. The co-construction modes we identify originated from detailed analyses of videotaped lessons and protocols. We focus in particular on a strategy called the Competition Mode, where the teacher promotes students to contribute to a discussion with ideas that are contradictory to each other. We argue that the presence of contradictory ideas can be productive in fostering dissonance and reasoning in students. Other co-construction modes, such as the Disconfirmation Mode and the Accretion Mode, are also identified. Each individual mode of interaction is represented diagrammatically. The diagrams are then combined to represent the effects of the rich exchange of ideas between the teacher and the students. We argue that the teacher in this study played a key role in the overall co-construction process. She constantly diagnosed the students' ideas and encouraged them to evaluate and modify them accordingly. Making teachers aware of these modes may provide important strategies for fostering model construction.

A medium size teacher-student interaction pattern, called the Competition Mode, was observed in a middle school classroom during model-based teaching and learning instruction. By medium sized, we mean that it does not take place within a single teacher student exchange but over many such exchanges; it usually occurs within a single class period. We will first define several of the modes in a theoretical way. Then several classroom episodes will be analyzed using the mode concepts developed.

The Competition Mode is defined as the process by which the students display or express to the class two or more competing ideas at a time, providing an opportunity for comparisons (and therefore dissonance) before closure is reached on an idea (such as models M1 and M2 shown in Fig. 7.1). For example, when students were asked to draw models of the structure of the throat before they had studied this topic, several different models were produced, including those with and without connections to the nose, and with one tube going downward vs. two tubes. By the

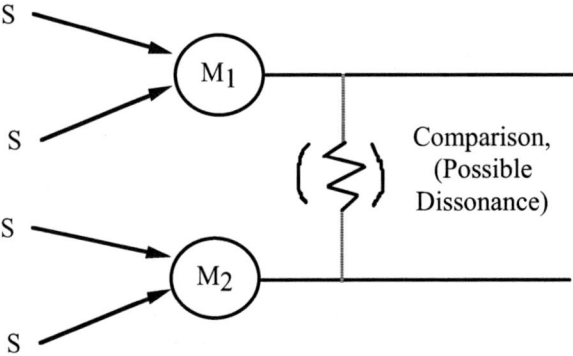

Fig. 7.1 Model competition

Competition Mode, we do not mean that the teacher fostered in the class a "competition for a prize." Instead, the teacher used discussions for supporting the students in comparing and making choices among the competing models. Early studies that describe competing ideas in student discussions are reported in Minstrell (1982) and Osborne and Wittrock (1983); here we attempt to provide an explicit definition above so that it can be recognized by others, differentiated from other modes, and examined in more depth for substrategies (Nunez-Oviedo, Clement, & Rea-Ramirez, 2003).

Other co-construction modes, such as the Disconfirmation Mode and the Accretion Mode, are also present in the selected episodes and will be discussed because they are closely connected to the Competition Mode. For example, the comparisons during the Competition Mode can naturally lead to the Disconfirmation Mode in which one of the models is criticized and drops out of the discussion (Fig. 7.2). In many cases, the criticisms can be elicited from the students. Dissonance (represented by the jagged line) can promote the evaluation process. The dissonance can be the result of the teacher or the students using a discrepant question, a thought experiment, negative feedback, or other techniques. As a product of the evaluation process a competing model that is not compatible with the target concept can be disconfirmed in the sense that it does not reappear in the classroom discussions. Thus, we call this teacher-student interaction pattern the Disconfirmation Mode (Fig. 7.2). (In using this term, we do not necessarily imply that the idea has completely disappeared from the students' minds. However, since such ideas did not reemerge in the discussions there is some evidence that the students considered them to be disconfirmed).

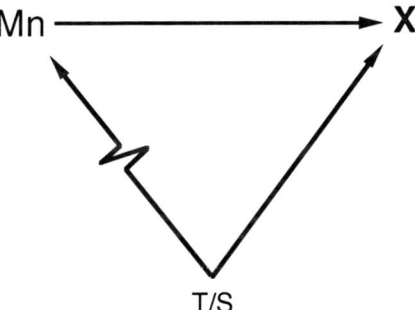

Fig. 7.2 Disconfirmation mode

In the Accretion Mode the teacher encourages the students to generate one small element of the model at a time. As Fig. 7.3 shows, the Accretion Mode occurs when the teacher asks students (S) a question (Q) and the students answer the question with an element that is compatible with the developing model (e.g., Sa, Sb or Sc). In response, the teacher evaluates the students' answer by providing positive feedback (PF). Thus, this teacher-student interaction pattern consists of a teacher's question (Q), student's answer (Sn), and a teacher's positive feedback (PF). The teacher provides the students with positive feedback by using words such as "good," "okay" or by immediately asking the next question. As a consequence of this pattern (Q-Sn-PF) there is a selective accretion of elements of the model (+abc) that are compatible with the target of the lesson. By asking a series of leading questions about easy ideas that students can infer, the teacher helps the students in putting together a string of small model elements where each student contributes with a piece that is essentially correct. At first sight, the Accretion Mode may look like "recitation" but it is quite different. The essential difference is that in the Accretion Mode, in most of the cases, the teacher has not yet taught the scientific model to the students. Therefore the teacher helps the students to use their prior knowledge in inventing ideas to piece together a new model. Recitation, on the other hand, refers to simply having students recall previously learned material.

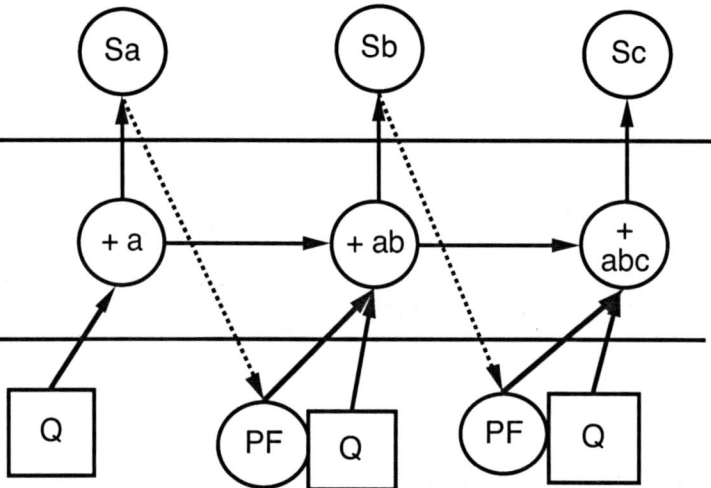

Fig. 7.3 Accretion mode

7.2 Coordinating the Co-construction Modes Above

We have combined the diagrams of the Competition Mode, the Disconfirmation Mode, and the Accretion Mode to explain three instructional sequences. The overall goal is to illustrate the resulting learning trajectory or learning pathway during model based teaching and learning instruction (Figs. 7.4, 7.5, and 7.10). The task, however, is complex and it is necessary to consider the organization of the scientific model that needs to be taught.

A model is made of individual elements or model elements that work together to explain how a certain structure or phenomenon occurs. For example, one can look at the structure of one cell as an entire model. Within the cell there is the nucleus that contains chromatin. Chromatin is made of DNA and forms the chromosomes. DNA is made of nucleotides each of which is composed by one sugar, one phosphate group, and one nitrogen base. The sugar is called deoxyribose and contains Carbon, Hydrogen, and Oxygen atoms. The cellular model contains embedded model elements that correspond to different levels of explanation. It is necessary to discuss these individually in order to understand the overall picture of this model.

Because an entire model is made of embedded model elements, we have found in many places teacher-student interaction patterns, that are also embedded while the teacher promotes the disconfirmation of a model element. For example, the Accretion Mode may occur as an "embedded strategy" or as a "sub-process" of the Disconfirmation Mode. Even though the Accretion Mode is applied to a model element, and the Disconfirmation Mode is applied to a complete model, we will not write the word "model" or "element" next to the name of the mode. We have taken this course in order to have simpler descriptions of the episodes.

7.2.1 First Example

The first example illustrates the way the teacher supports the students in examining a model element. This example was recorded during the teaching of a digestion unit where the teacher introduced the idea that fat is the secondary source of energy for the body. The episode began when the teacher and the students agreed that carbohydrates, fats, and proteins are sources of energy for the body. They also agreed that carbohydrates are the primary source of energy for the body. The teacher then asked the students to discuss within their groups and report to the class whether proteins or fat were the secondary sources of energy for the body. She found that half of the groups held the idea that fats were the secondary source of energy while the other half of the groups said that proteins were the secondary source of energy for the body. To accomplish this, the teacher used the Competition Mode (Figs. 7.4 and 7.5).

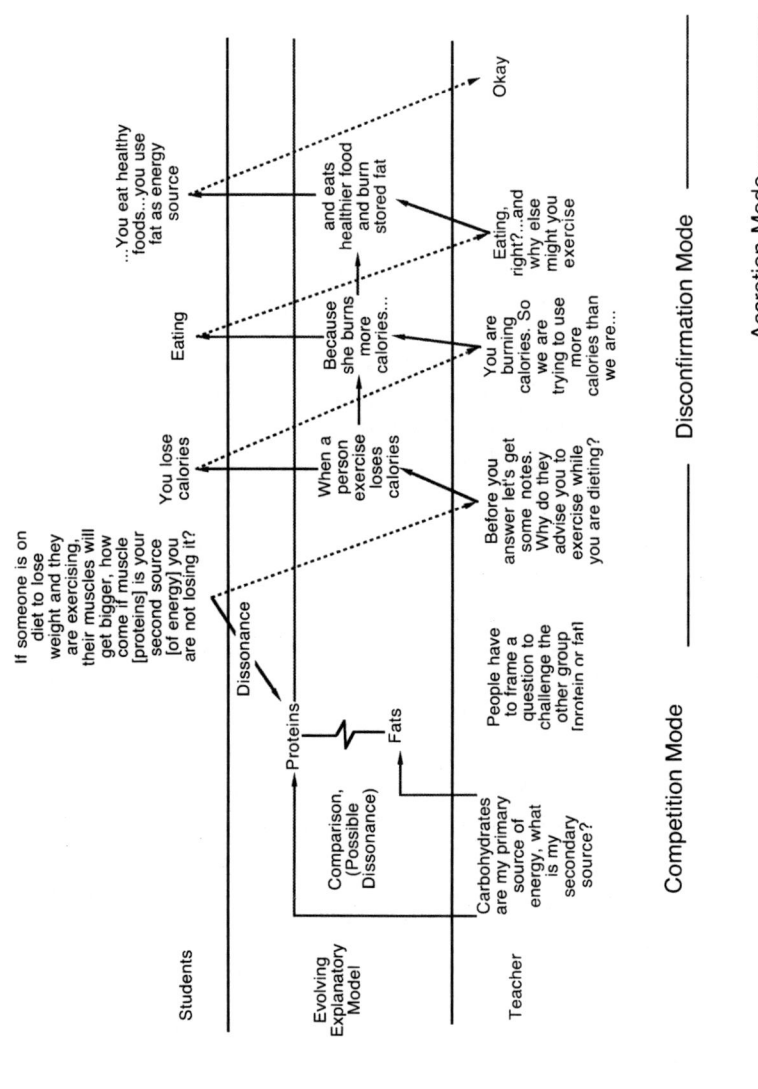

Fig. 7.4 Coordination of the co-construction modes to build and evaluate an individual model element: first example

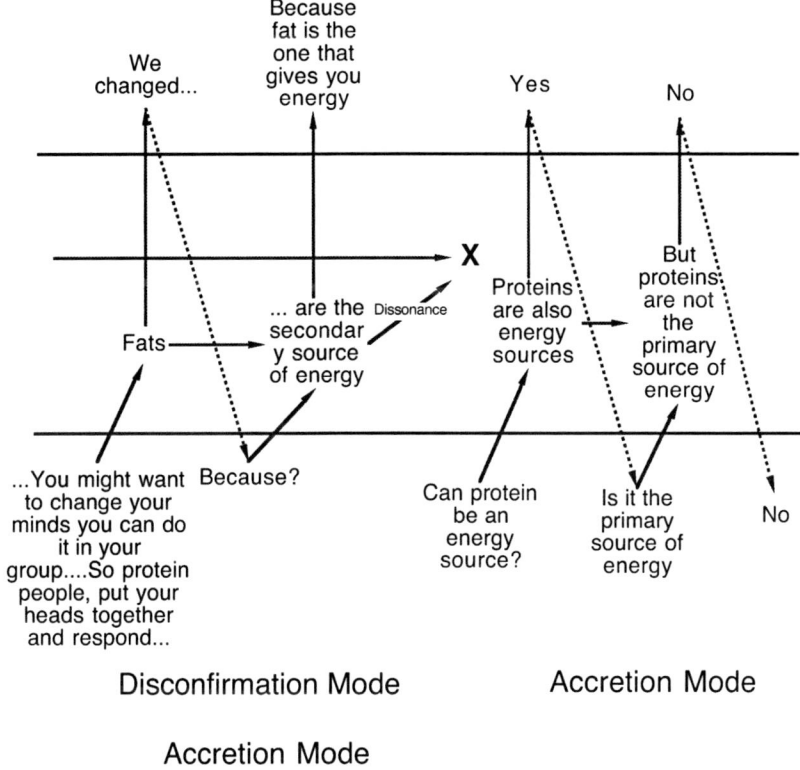

Fig. 7.5 A close look at the disconfirmation and accretion segment

The teacher then fostered the Disconfirmation Mode in which the students dropped proteins as a secondary source of energy for the body. In order to disconfirm this idea, the teacher asked the groups to challenge the opposing statement by asking a question. However, only a group that believed that fats are the secondary source of energy for the body was able to invent a viable discrepant question for the challenge. A student asked, "If someone is on a diet to loose weight, um and they are exercising, usually their muscles will get bigger than if they loose fat. How come if, um, muscle is your second source, how come you are not loosing it?" It is interesting to note that the teacher supported the students in asking a discrepant question to introduce dissonance in their classmates.

Before allowing other students to answer the question posed by the student above, the teacher used the Accretion Mode to discuss with the class the concepts of calories, dieting and exercising while loosing weight (see Fig. 7.5) in order to make sure that they understood the question. In other words, the teacher used the Accretion Mode as a subprocess of the Disconfirmation Mode. It is important to mention that she had not explicitly taught to the students about diet or calories throughout the unit. Therefore, the teacher helped the students put together a number of model elements that are essentially correct by using their prior knowledge about the topic. As a result of the discussion, the class came to the conclusion that, because the person was exercising, she was going to loose more calories than those she was ingesting and that those calories would come from her body fat. After reflection, the students in the protein group said that they had changed their vote to "fat". Thus there is some evidence that a number of students had changed their minds to a more correct view during this lesson.

7.2.2 Second Example

A second example also illustrates the way the teacher supported the students in examining and discounting model elements. This episode was recorded during the teaching of a mitochondria unit. The main objective of this episode was to help students understand how the energy stored in the bonds of the glucose molecule is transferred into an ATP molecule. The scientific model held that this process occurs within the mitochondrion.

The episode began when the teacher asked the students to work individually to draw and label a cell. She also asked them to indicate in which organelle the energy contained in the glucose molecule would be transferred to the ATP molecules. The teacher also told the students that it was "okay" if they drew and labeled more than one place within the cell as their hypothesized energy transfer location. The teacher then asked the students to share, compare, and discuss their ideas within their small group and to write down the group's answer on a dry-erase board, or white board. Once the students agreed about the group answer, the teacher invited the small groups to report their ideas to the large group.

Even though the students had learned previously about the individual functions of the cell organelles, the groups reported, in total, five different places within the cell where the energy transfer might occur: the cell membrane, the Golgi body, the endoplasmic reticulum (ER), the nucleus, and the mitochondria. In other words, the students reported model elements that were not compatible with each other and many that were not compatible with the target model.

In the first episode of the Disconfirmation Mode, the teacher supported the students in discounting the cell membrane as the place where energy transfer occurs within the cells (Fig. 7.5). It should be noted that during the Disconfirmation Mode the teacher does not simply tell students that a perceived model is incorrect since this has been found to be ineffective in helping students modify their mental models. The teacher in this instance began by asking the students about the function of the cell membrane. A student answered that the function of the cell membrane was "to let things in and out" and the teacher said "okay." The teacher then asked the class a discrepant question. She asked whether the cell membrane is the area of the cell in which the energy transfer occurs. Many students answered "no" and the teacher agreed with them by saying that the function of the cell membrane is to protect the cell. The teacher allowed the students to disconfirm the cell membrane as the place for energy transfer by asking them whether they wanted to get rid of the cell membrane. Several students expressed agreement and a student said loudly, "take it out" and the teacher repeated the student's response with an affirmative tone. As a result of this conversation, we hypothesized that other students also dismissed the idea that energy transfer occurs within the cell membrane.

In the second episode of the Disconfirmation Mode, the teacher helped the students discount the Golgi body and the ER as possible places for energy transfer from the glucose to the ATP molecule (Fig. 7.7). In order to do this, the teacher promoted several episodes of the Accretion Mode that were sub-processes of the Disconfirmation Mode. The teacher asked the students a string of questions to help them recall one model element at a time about the function of the Golgi body until the information was complete. The teacher did the same with the ER's function. Based on the information provided by the students, the teacher asked them whether the Golgi body and the ER were connected to transfer of energy contained in the glucose. Most of the class answered "no" and the teacher repeated the answer with an affirmative tone. She asked the students whether they wanted to continue discussing these organelles as places for the energy transfer in the cell and most of them answered "no." As a result of this conversation, we hypothesized that most of the students also dropped the idea that energy transfer occurs within the Golgi body and ER.

In the third episode of the Disconfirmation Mode, the teacher focused in discounting from further consideration the nucleus as the place where energy transfer occurs within the cell (see Fig. 7.7). However, even before the teacher asked any questions a student concluded that the nucleus was not the place for energy transfer. In spite of that, the teacher continued asked the students why then the nucleus is not in charge of the energy transfer within the cell; a student said that the nucleus is in charge because

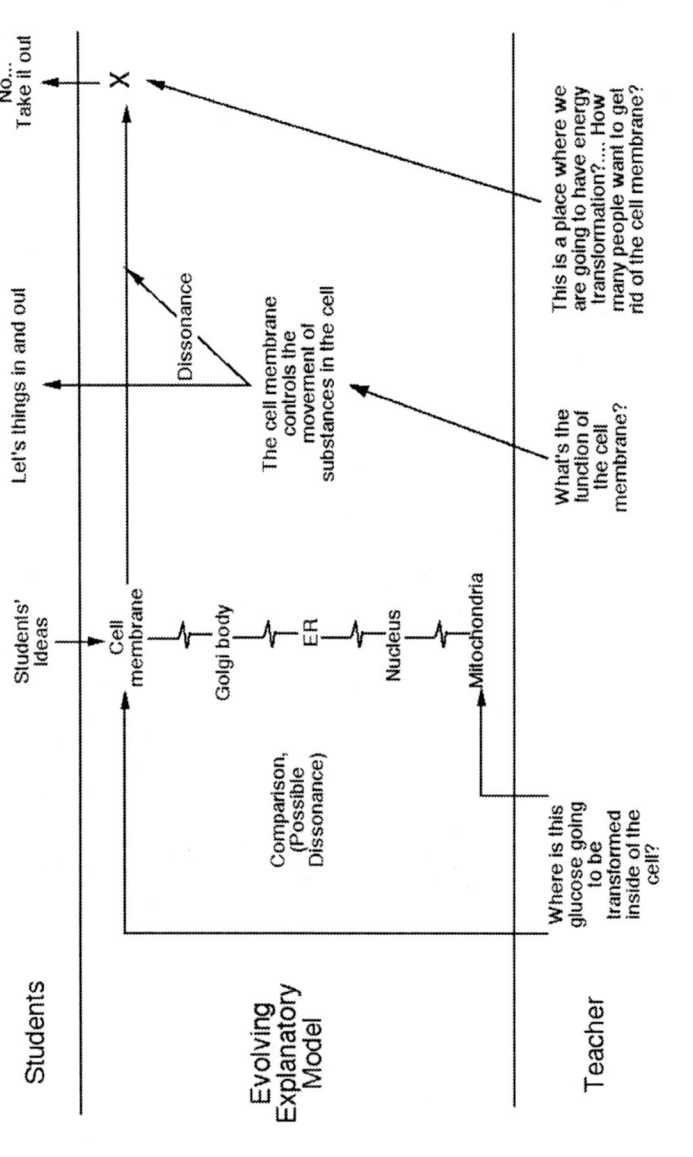

Fig. 7.6 Coordination of the co-construction modes to build and evaluate an individual model element: second example

Competition Strategy and Other Modes   127

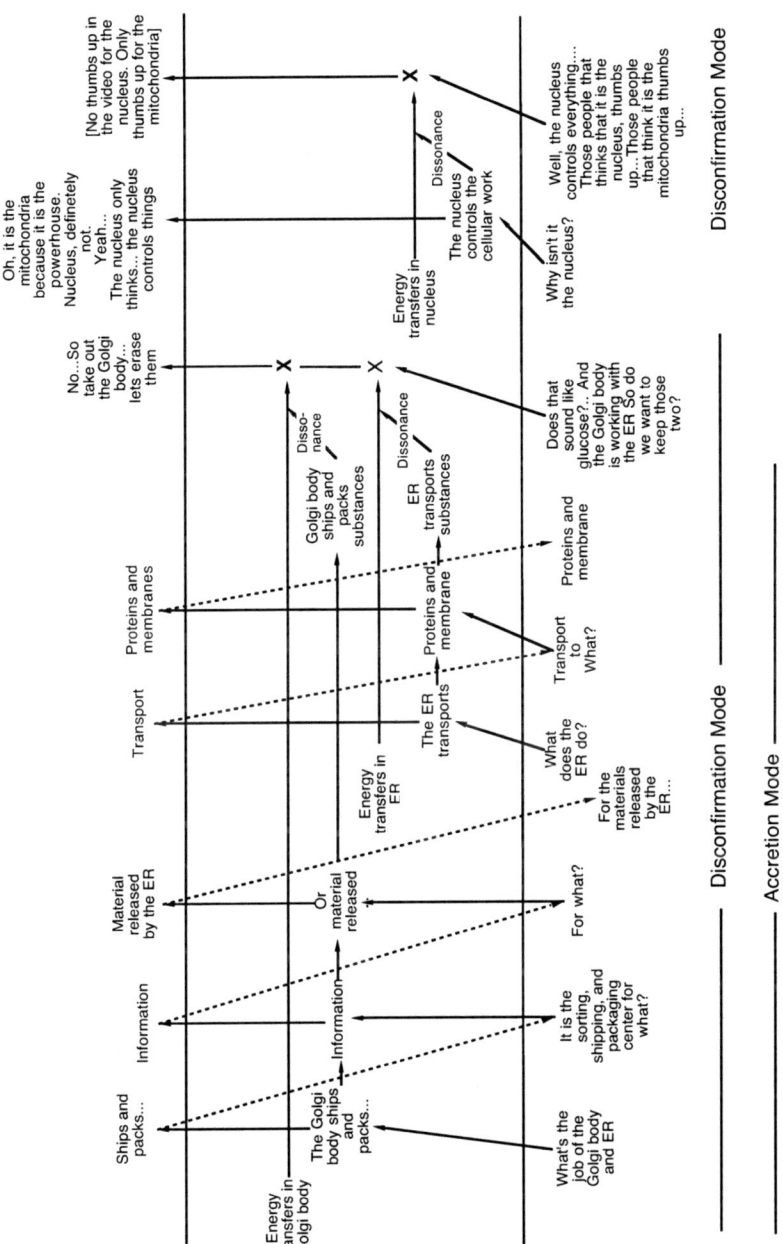

Fig. 7.7 Detailed interaction and modes in second example

it "thinks." The teacher rephrased the student's answer by saying that the nucleus controls the cellular work. The teacher encouraged the students to vote whether the nucleus or the mitochondrion was the place in the cell where energy is transferred within the cell. She asked them to use their thumbs up or down to express their opinion. Based on the video observation, all of the students put their thumbs up for voting the mitochondrion.

In summary, this example shows how the teacher supported the students in removing from further consideration the cell membrane, the Golgi body, the ER, and the nucleus as the cell organelles where energy is transferred from the glucose to ATP molecules. In this episode we observed the teacher fostering an episode of the Competition Mode. The teacher also fostered three episodes of the Disconfirmation Mode to discount the cell membrane, the Golgi body-ER, and the nucleus, from further consideration (Fig. 7.6). In each Disconfirmation Mode, the teacher used discrepant questions to promote student's reasoning. In order to disconfirm the Golgi body and the ER, the teacher conducted a small episode of the Accretion Mode that was a sub-process of the Disconfirmation Mode (Fig. 7.7). This example provides evidence that the teacher supported the students in conducting successive evaluation cycles of model elements until settling on an idea compatible with the target model.

7.2.3 Third Example

The third example illustrates the way the teacher supported students in examining and discounting model elements and entire models. This episode was recorded during a lesson on the structure of the throat and it was included in the unit on digestion located at the beginning of the curriculum. It lasted approximately 45 minutes. This is a very complex instructional sequence that took place before the students had been taught about the target concept. The description of the sequence includes embedded co-construction modes or layers of sub-processes.

The structure of the throat is a complex topic for students (Fig. 7.6). The scientific model describes the throat as an intricate place located between the mouth and the esophagus that leads the food into the stomach. The trachea is a tube located in front of the esophagus that emerges from the larynx and allows air to travel to the lungs. This passage is usually open and it only closes when a person swallows. Several structures converge at the throat or pharynx including the mouth, nose opening, and the entrances of the esophagus and larynx. The ears are also connected to this place through the Eustachian tubes that lead into the throat from the middle ear. A flap of skin called the epiglottis is situated above the larynx. It covers the entrance and directs the food back to the esophagus when a person swallows.

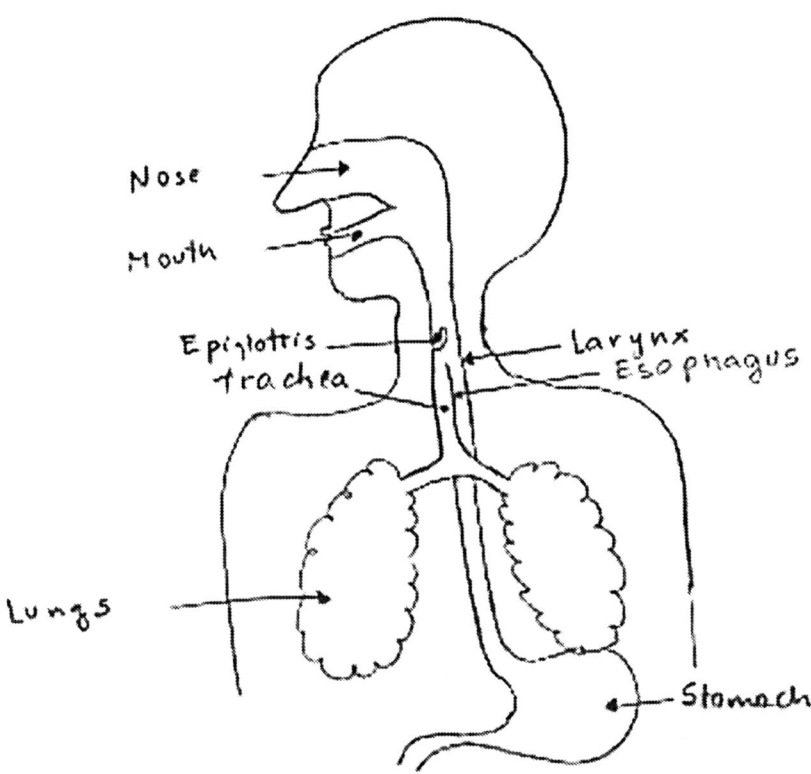

Fig. 7.8 Structure of the throat

The students had initial ideas about the throat structure that were markedly different from the target model. The teacher detected these conceptions by asking the students to draw their ideas about the throat individually and then to share them within a small group. The teacher then asked the students to create a group consensus drawing on a small whiteboard. The teacher asked the students to compare each other's group drawings to determine similarities and differences between them. To do this, the teacher organized this activity in such a way that one member of each small group stayed at the table explaining the group's drawing to the incoming students while the other three members of the group went to examine other group drawings ("three stray, one stays.") The teacher then asked the students to identify the models that were markedly different from each other, and they selected the three models described below.

In model 1, the students drew a tube going from the mouth and another tube going from the nose that joined together in the back of mouth. Further down, this tube split into two tubes going to the lungs. (see Fig. 7.9)

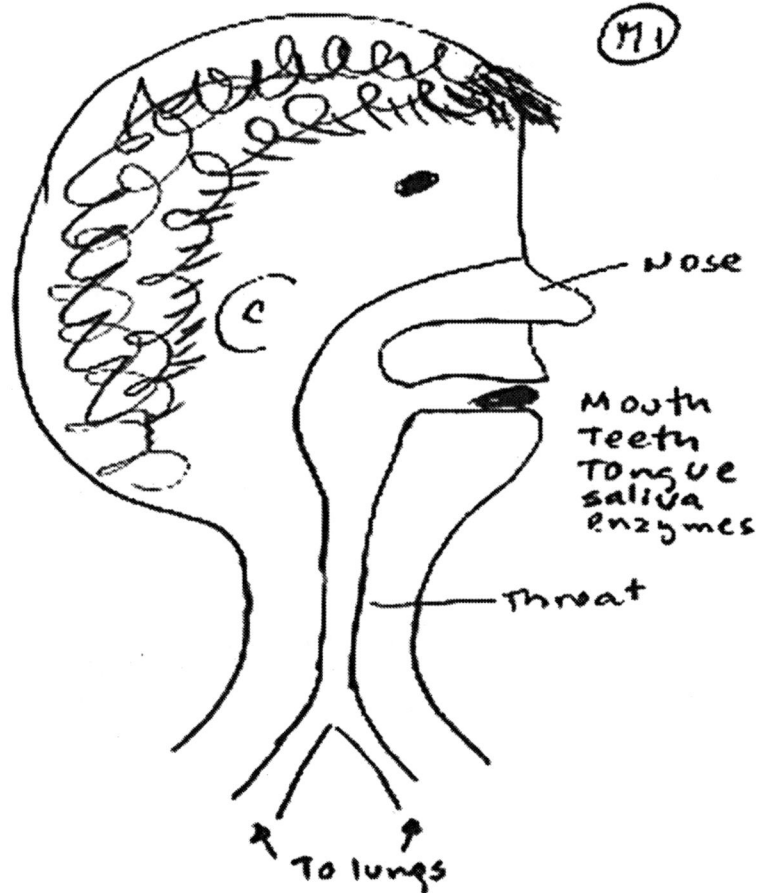

Fig. 7.9 Model 1

In model 2, the students drew one tube going from the nose down to the lungs and another tube going from the mouth down to the stomach (see Fig. 7.10).

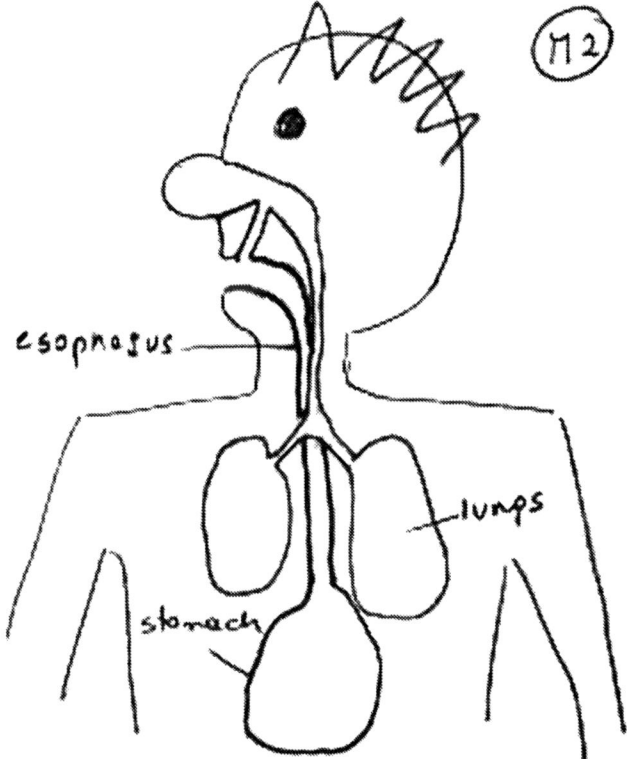

Fig. 7.10 Model 2

In model 3, the students drew a tube going from the mouth, another tube going from the nose that joined together in the back of the mouth, and one tube that was going down from the mouth (see Fig. 7.11).

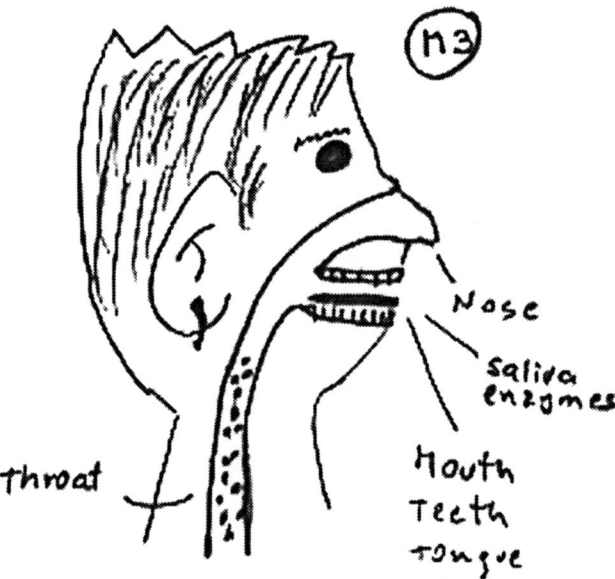

Fig. 7.11 Model 3

We observed that to foster conceptual change in the students, the teacher coordinated 3 episodes of the Competition Mode, 2 episodes of the Disconfirmation Mode, and 2 episodes of the Accretion Mode. We also observed that the teacher used the Accretion Mode as a sub-process of the Disconfirmation Mode and that the disconfirmation of a model element was a sub-process of the disconfirmation of an entire model.

The first episode of the Competition Mode occurred while students were examining each other's drawings (Fig. 7.12). The teacher asked the students to select three drawings (models 1, 2, and 3) to compare. None of these models was in agreement with the target concept and the teacher first encouraged the students to disconfirm those models (1 and 2) that were least compatible with the target concept by originating two episodes of the Disconfirmation Mode (Fig. 7.12). The first episode of the Disconfirmation Mode occurred when the teacher encouraged the students to discount model 1 from further consideration (Fig. 7.12). The teacher took the whiteboard in her hands and described the diagramed model aloud. She said that the drawing showed a tube going from the mouth and another tube going from the nose that joined together in the back of mouth. The drawing also showed one tube that split off into two tubes below the throat. The teacher asked the group that had drawn the diagram where these two tubes were going and a member said that these two tubes were going to the lungs. The teacher then asked the class a discrepant question, asking whether food was going into our lungs and a student answered "no". The teacher then added that model 1 was taking care of the air but not the food. The process of discounting model 1 appeared to have been successful because none of the members of the class brought model 1 back to the discussion in the rest of the lesson.

The second episode of the Competition Mode took place when the teacher encouraged the students to compare models 2 and 3 (Fig. 7.13). She first asked the students to show to the class the two whiteboards containing the drawings and described them aloud. She then asked the class whether they could see the difference between these two pictures and a student said loudly "yeah." She also encouraged the students to come up with a question to challenge these models but her efforts were not successful. Instead, a student asked the teacher which of the models shown in front of the class was right. She told him that she was not going to answer that question. The teacher then used the Accretion Mode in attempting to gather evidence for supporting either model 2 or 3. They discussed that the esophagus and the stomach are connected, and a student pointed out that model 2 implies that a person could breathe and swallow at the same time because it has two tubes.

The third episode of the Competition Mode took place when the teacher encouraged the students to select model 2 or 3 by voting. She asked the students "what do you think is the best or more efficient model" (Fig. 7.13). The voting, however, was for the second time unsuccessful in discounting one of the models because it resulted in a tie. The teacher then made a new attempt at disconfirming one of the models.

The second episode of the Disconfirmation Mode took place when teacher supported the students in discounting model 2 from further consideration (Fig. 7.13). The teacher used a discrepant question based on the statement made by a student. The teacher asked the class whether a person could swallow and breathe at the same time and the class gave to the teacher a resounding "no." The teacher complemented the students' idea and asked what occurs when a person swallows and breathes at the same time; the students answered that the person would choke.

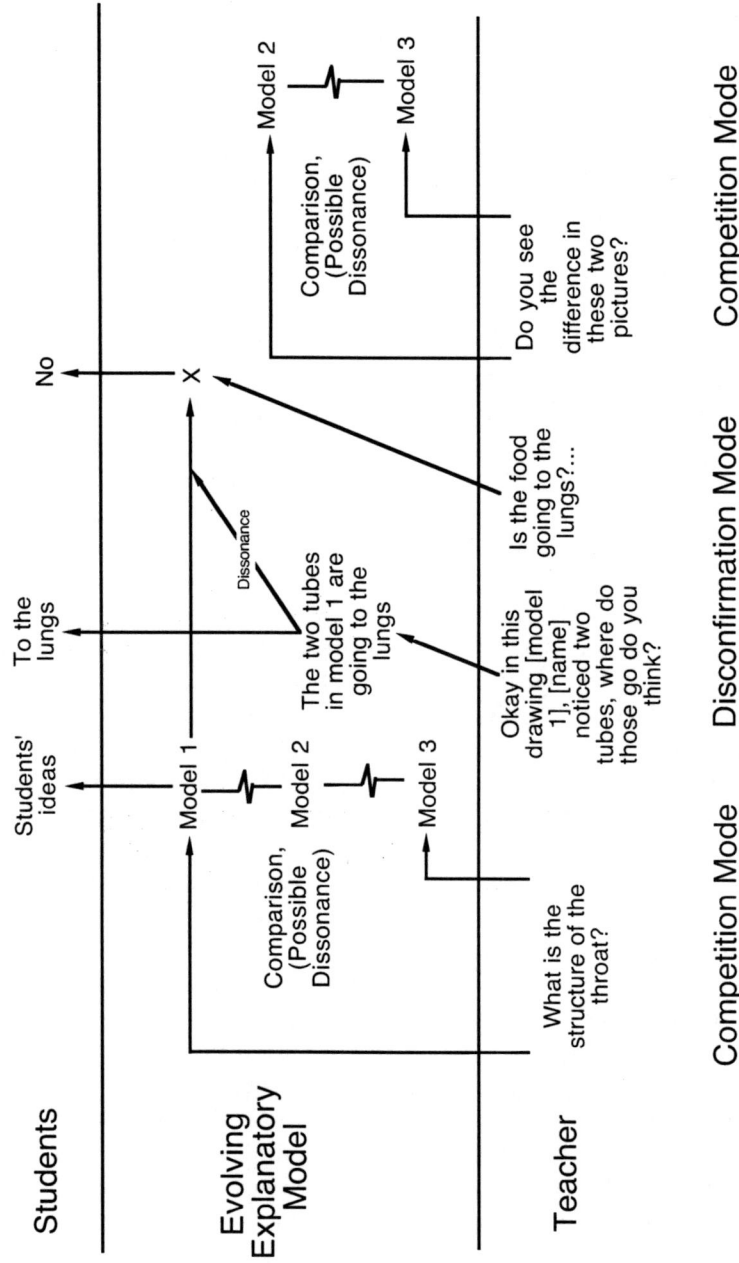

Fig. 7.12 Coordination of the co-construction modes to build and evaluate entire models

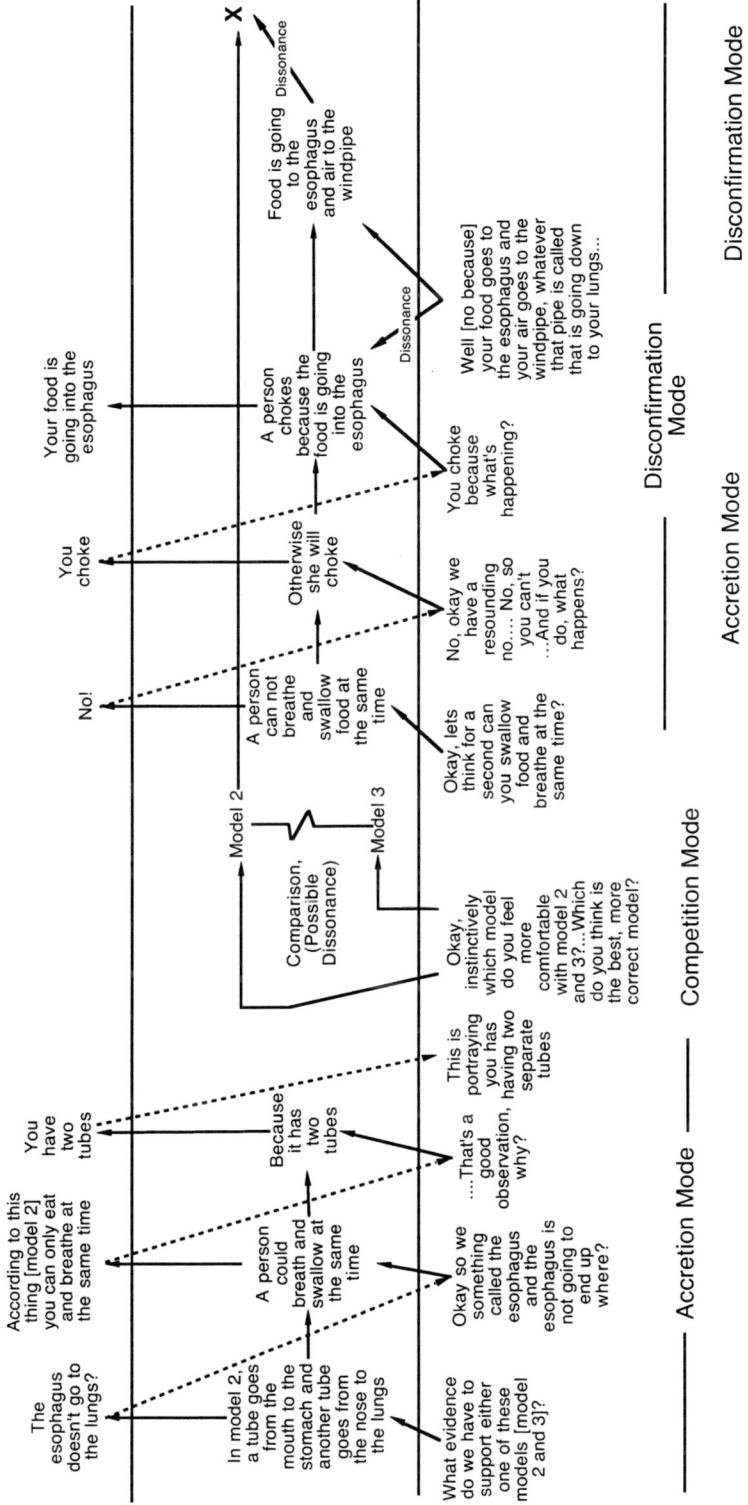

Fig. 7.13 Detailed analysis of co-construction of the throat

The teacher then conducted a second episode of the Accretion Mode as a sub-process of the Disconfirmation Mode. She asked about the cause of choking. At this point a student tried to explain choking by saying that food "goes into the esophagus". Perhaps the student meant, "should go into the esophagus. Perhaps the student meant, "that food goes into the windpipe." The teacher then clarified the reason for the inadequacy of Model 2. The teacher told the students that normally "food" goes into the esophagus and that the "air" goes into the windpipe that is going into the lungs. Thus the teacher and the students had found a reason for discounting model 2 from further consideration.

By the end of the period only model 3 had survived the teacher's and the student's close evaluation. However, this model was also far from the target model and the teacher supported the students in evaluating and modifying model 3, originating a teacher-student interaction pattern called Model Evolution. In the Model Evolution mode, one engages the students in evaluating and modifying a model repeatedly until one reaches the target model. That mode will be discussed in Chapter 10 of this book. The class made additions and improvements to Model 3 until they reached the target model.

7.3 Conclusion

Throughout this chapter we have described the strategies used by a successful middle school science teacher to support the process of building mental models in her class. We have interpreted the teacher's actions and students' learning processes by using our newly developed vocabulary that includes the Competition Mode, the Disconfirmation Mode, and the Accretion Mode, for comparing discounting or building up model elements and entire models. Our purpose has not been to pass judgment on whether the teacher's strategies were optimal; rather, it has been to identify and define new concepts for describing modes of teaching and learning in the classroom. In the examples discussed in this chapter, the teacher did not just tell the students that some of their model elements were wrong. Instead, she guided the students to conclude that some of their model elements were not viable and coordinated different co-construction modes to produce changes in the student's models.

The Competition Mode is a very interesting teacher-student interaction pattern because it feels very different from a lecture/presentation mode. Typically it would originate when the teacher asked the students to make an educated guess about the structure of a model by discussing their ideas within a small group and then reporting them in large group. We hypothesize that there are two purposes for this mode. The first purpose of the Competition Mode is to allow the teacher to detect and lay out the various student ideas with regard to a topic. The second purpose of the Competition Mode is to make the students aware that they have conflicting ideas regarding a topic, and to provide an opportunity to make comparisons and possibly to introduce dissonance. As was discussed in the three examples, many of the students' initial ideas were not compatible with the target model. The Competition Mode made the teacher aware of these ideas and helped her to design a plan to deal with them by using the Disconfirmation Mode along with building up ideas using the Accretion Mode (or via the Evolution Mode discussed in Chapter 10). However, using the Disconfirmation Mode in evaluating and discounting an idea is not a quick or "effort-free" strategy. On the contrary, the teacher often promoted student reasoning by inventing discrepant questions to foster dissonance.

On the other hand, the teacher used the Accretion Mode for stimulating students collectively to add new pieces to the model. It was observed that the teacher continued the Accretion Mode as long as the students continued adding pieces compatible with the target model. Use of the Accretion Mode ended when the students failed to provide the next correct piece of the model.

The process of encountering successive episodes of dissonance that lead to the evaluation and modification of preconceptions resembles Nerssesian's (1995) and Clement's (1989) descriptions of the way expert scientists reason about a domain. However, students may differ from experts in their level of preparedness for evaluating models. The teacher discussed in these examples played a large role in examining students' ideas and providing them with small constraints or requests that produce dissonance and that encouraged them to examine their ideas and modify them accordingly. As we saw in example 3, sometimes students are able to play the predominant role in criticizing a proposed model, especially if prompted and given time to do so.

Accounting for all of the major cognitive strategies used in these transcripts pushed us to define several basic teacher-students interaction modes. Because they have more specific cognitive purposes for model construction, these modes are different from the more general strategies identified by van Zee and Minstrell (1997) for facilitating and keeping large group discussions going by encouraging students to contribute. This chapter should be read in conjunction with Chapter 10, which documents another essential mode, the Model Evolution Mode, used by the same teacher. That mode builds up the new model in a more creative way than the Accretion Mode discussed here. Together these modes constitute a set of higher order strategies for fostering model construction.

References

Clement, J. (1989). Learning via model construction and criticism. In G. Glover, R. Ronning, & C. Reynolds (Eds.), *Handbook of creativity: Assessment, theory and research*, (pp. 341–381). New York: Plenum.

Minstrell, J. (January 1982). Explaining the "at rest" condition of an object. *The Physics Teacher*, 10–14.

Nersessian, N. J. (1995). Should physicists preach what they practice? Constructive modeling in doing and learning physics. *Science & Education, 4*, 203–226.

Nunez-Oviedo, M. C., Clement, J., & Rea-Ramirez, M. A. (2003). Model competition and other modes of large group discussions for model based learning. Proceedings of NARST.

Osborne, R., & Wittrock, M. (1983). Learning science: A generative process. *Science Education, 67*(4), 489–508.

van Zee, E., & Minstrell, J. (1997a). Using questioning to guide student thinking. *The Journal of the Learning Sciences, 6*(2), 227–269.

Chapter 8
What If Scenarios for Testing Student Models in Chemistry

Samia Khan

University of British Columbia

8.1 Introduction

This chapter describes how teachers can engage students in testing their models using a teaching strategy known as *what if scenarios*. What if scenarios involve speculating on or changing one or more of the parameters or factors associated with an original model and observing the effects of the change. For example, in a "what if" scenario, one can ask what would happen to Galapagos Island finches if a drought destroying the small, soft seeds finches eat occurred over a period of time (Reiser, Tabak, Sandoval, Smith, Steinmuller & Leone, 2001)? By observing, or speculating on, the possible effects of the change, assumptions about a model can be uncovered, boundaries of the model can be ascertained, and implications of running a model can be explored. To follow the example, a documented effect of the climatic disturbance was that smaller-beaked finches died out more than they had in the past and in larger proportions to larger-beaked finches. One's assumptions about the process of natural selection can be uncovered by examining the effects of climatic disturbance to changes in the finch population, and if relative drought conditions are "run" over a period of a few hundred generations of finches on the Galapagos Island, the implications for contemporary evolution of organisms could be further explored.

Fields such as economics, urban planning, and meteorology, have historically employed what if scenarios to extrapolate how a proposed model changes over time as a method to learn more about markets, plan transportation, or prepare for natural disasters. What if scenarios can also be employed to learn science. Drawing upon an in-depth analysis of a teaching episode with a class of introductory-level university chemistry students, the use of what if scenarios by a chemistry teacher are examined. In this teaching episode, the teacher guides students to re-examine a model that they constructed and expressed earlier in the lesson. The teacher engages students in this re-examination of their original model by asking students a *what if* question that provokes them to consider changing a single parameter or condition associated with the original model they constructed. The teacher then suggests that students test their model under this new condition. Testing occurs in this case in a virtual environment generated by web-based java applets. The technique of utilizing digital technology for what if scenarios should have many potential applications because numerous applets are housed in various digital repositories on the web and are accessible to teachers.

8.2 Model-based Learning and What if Scenarios

Model-based learning involves constructing and transforming models. A possible way to transform models is to re-examine an original model under novel conditions. What if scenarios can be used to provoke such a re-examination because what if scenarios require testing of a model under novel conditions. Students may be motivated to reconsider aspects of an original model upon testing, especially when the student discovers that the original model cannot adequately explain the outcomes of the what if scenario. Reconsidering aspects of an original model can prompt significant changes in students' conceptions of a phenomena. In the teaching episode below, the goal of the teacher was to foster understanding of the atomic model. During this episode, the students constructed a revised model of the atom from the models they were presented with in high school. In this case, the teacher asked the students several what if questions. These what if questions appeared to set the stage for a re-examination of the atomic model. Re-examination of the atomic model provoked discussion about the model among the students and resulted in a new atomic model that was used later to explain nuclear stability.

8.3 Models of the Atom Presented in High School

In high school chemistry, the curriculum usually includes various models of the atom. These models usually include those based on historical experiments by Dalton, Thomson, Rutherford, Bohr and Schrödinger. In the

1800's, Dalton looked for patterns in chemical reactions and suggested that the atom is a small solid sphere that could not be divided, like a billiard ball. By 1904, Thomson had discovered the electron and constructed a model of an atom that was a sphere of positive charge with the new particles, electrons, embedded in it. The Thomson model was not like the billiard ball model; it was sometimes referred to as a "plum pudding" model of the atom because the electrons could be described as being like the plums scattered throughout and suspended within a positively charged electric field (the "pudding"). The negative charge of the electron and the positive charge of the "pudding" would "cancel" one another out leaving the charge of the entire atom at zero. In the "plum pudding" model, Thomson had no empty space. Rutherford's experimental work that followed later showed that: certain heavy atoms spontaneously decay into slightly lighter, and chemically different atoms; the atom is largely empty space, and the nucleus is very small, positively charged, and contains most of the mass of the atom. Rutherford's findings led to the abandonment of the billiard ball model and the plum pudding model of the atom. Building on Rutherford's work, a "planetary model" of the atom emerged with Bohr's experiments. The "planetary model" involved electrons surrounding the nucleus in specific energy levels. The electrons were compared by some with planets traveling in a ring around the nucleus that was like the sun. Erwin Schrödinger built upon the work of Bohr developing the probability function for the Hydrogen atom. The probability function describes a cloud-like region where the electron is likely to be found rather than following a circular pathway like an orbit. If the collection of electron traces is assembled, it begins to resemble a cloud. The "cloud" represents a history of where the electron has probably been and where it is likely to be going. The "billiard ball", "plum pudding", "planetary", and "cloud" models are some of the models of the atom that students are typically introduced to in high school chemistry.

The electrostatic force is the force arising between two static electric charges. This force is proportional to the product of the electric charges, and inversely proportional to the distance between the charges. These relationships are represented quantitatively as Coulomb's Law. The electrostatic force is one type of electromagnetic force that allows electrons to stay with nuclei, atoms to bond and form molecules, and more generally, creates the substances we interact with all of the time. Understanding the electrostatic force is essential to many discussions of the model of the atom.

8.4 Testing a Model Using a What if Scenario: The Case of Nuclear Stability

Many students from high school have learned models of the atom that resemble the planetary model. The teaching episode below illustrates how a teacher in this study helped students test the parameters of planetary and cloud models using a series of what if scenarios. After encouraging students to share their models of the atom, a teacher can enter into a phase of teaching where a student's model of the atom is ready to be explored further and potentially transformed. This chapter episode focuses on the second part of this interaction. Web-based java applets can support exploration and transformation of students' models by providing a virtual environment for students to test their models. In this teaching episode, students' models of the atom were explored further using a web-based java applet (Coulomb's Law simulation). The Coulomb's Law simulation provided a virtual environment for students to change parameters such as charge, magnitude of charge, and distance between two depictions of charged particles. The screen shot below is the Coulomb's Law simulation portraying two charged particles, ions that can be placed a variable distance from each other.

Fig. 8.1 Coulomb's law simulation

In this simulation, students can change the distance between the ions by grabbing the ion with their mouse or touchpad and dragging it further or closer to the stationary ion on the screen. Students can change the charge on both ions either to positive or negative, or increase or decrease the same charge by using the up or down arrow keys on a keyboard. The arrows on the ions visible on the screen represent the force of attraction. The arrows increase or decrease in size as distance or charge increases or decreases. Students can also view changes in force and distance as numerical values.

Returning to the teaching episode, the teacher (T) engaged students (S1, S2, S3) in testing the parameters of electrostatic force and distance between two hypothetical charged particles using the Coulomb's Law simulation, as depicted in a transcript of a classroom discussion below.

T We can do experiments to manipulate matter. Our goal is to bring everything back to the structures of atoms. Atoms have three particles. We will use simulations to get a gut feeling of the relationships in chemistry. Go to the Coulomb's law simulation.

(T Demonstrates the Coulomb's Law simulation and how it works).

T Play, observe, write down what you observe, come up with the rules. Who can tell me the relationship between distance and the electrostatic force? *What would happen if* the distance was double?

(Students are observed changing the distance parameter in the simulation).

S1 As distance increases, the force gets smaller.
S2 It's d^2!
The teacher now asked students what *would happen if* the magnitude of the charge changed:
S3 It changes in increments.
T What do you see?
S1 When we go from –1 to –2, the force doubles.

Proceeding with additional what if scenarios in the series, the teacher asked students *what would happen if* the magnitude of the charge changed from –2 to –3 or from –1 to –3? Students' co-constructed with their teacher that the larger the charge, the stronger the electrostatic force, and the larger the distance, the weaker the electrostatic force. The relationship was also represented quantitatively as force being directly proportional to charge 1 multiplied by charge 2 over d squared.

Students were then asked a series of what if questions about a sequence of incremental changes to a variable; in this case, a magnitude of charge incrementally changed from –1 to –2 to –3. These teacher-student interactions led to the generation of two semi-quantitative relationships (a direction of change relationship between two ordinal variables such as "increasing A causes an increase in B") and one quantitative relationship regarding distance between charged particles and force. The what if scenario that proceeded incrementally afforded students with opportunities to explore the relationship between the distance between protons and electrons, magnitude of charges, and force of attraction between these particles, all relevant to the model of the atom.

After students expressed their understanding of relationships among particles, the teacher then asked students to re-examine their model of an atom. The teacher drew a picture of a magnesium atom and asked what is wrong with this picture based on what was learned from the previous data set.

> T Draws a cloud picture of ^{24}Mg. *What's wrong with this picture based on Coulomb's Law?*
> Students respond by reexamining their original planetary and cloud models:
>
> S1 *Why don't electrons pull into the protons?*
> S2 *Is the distance between the electron cloud and nucleus set?*
> S1 We learnt it as rings, remember?
> T What doesn't make sense?
> S6 Some electrons should be at different places like a p orbital.
> S7 *Why don't electrons collapse into the nucleus?*
> T Electrons are always trying to get closer to the nuclei. Always.
> S8 What is between the cloud and the nucleus?
> T Mostly a vacuum. *Another glaring problem!*
> S9 *Why do all the protons stick together in the nucleus?*

The teacher concluded this particular learning episode with the idea of the "strong" force. The teacher explained that theoretical models of the atom based on Coulomb's Law can explain most experiments to incredible accuracy, but other fundamental forces also play important roles inside the nucleus. A special force was postulated to overcome the electric repulsion between protons in the nucleus, and because of its strength at short distances, it was "creatively" termed the "strong force".[1]

In terms of the student question regarding electrons collapsing into the nucleus, the teacher can return to historical models and extend the discussion to include quantum mechanics. As content background, in 1913, Bohr generated a "planetary" model of the atom where the electrons were limited to specific orbits around the nucleus. This model was based on classical physics; however, it was later discovered that this theoretical model would not hold up as orbiting electrons must crash into the nucleus due to a loss of energy from radiation caused by the movement of the electrons. Therefore, the atom could not be stable in Bohr's model. To get around this problem, quantum physics was introduced. In quantum theory, electrons are said to possess "quantized" or packet energy, which can only be released when they jump from the outer orbit (higher energy level) to the inner orbit (lower energy level). Thus, if the electron is at its lowest energy level, it has no lower level to which it can jump. This atomic configuration is stable and there is no chance for the electrons to collapse into the nucleus. Furthermore, in quantum theory, small particles like electrons possess particle-wave duality. Waves have special properties, such as interference that was first demonstrated by Young's slit experiment. Because of these wave properties, electrons cannot stay on circular orbitals but travel unpredictably around the nucleus.[2] Therefore, a "cloud" is drawn to show the region of space in which the particle it represents is located.[3]

At this point, the teacher can continue to discuss nuclear stability in terms of other forces such as: weak nuclear force, electromagnetic force and gravitational forces or, as was the case in this lesson, enter into a discussion on binding energies. As content background, nuclei are made up of protons and neutrons, but the mass of a nucleus is experimentally observed to be less than the sum of the individual masses of the protons and neutrons that constitute it. The theory is that this difference in mass is because it has been converted to nuclear binding energy that holds the nucleus together. As a measure of the strength of the strong force, this binding

---

[1] After the discovery of quarks, however, scientists realized that the force was actually acting upon the quarks and gluons making up the protons, not the protons themselves.

[2] Heisenberg's uncertainty principle states that the position and momentum of a particle cannot be determined simultaneously.

[3] Because of electrostatic forces between electrons, electrons cannot get too close to each other, which means an orbital can only hold a maximum of 2 electrons to avoid high repulsion forces. Some electrons may be close to the nucleus; others may be relatively further away.

energy can be calculated from the Einstein relationship $E=mc^2$. The greater the mass loss, the greater the nuclear stability, because the greater the nuclear binding energy. An interesting case to look at is ^{56}Fe. As students concluded in a discussion on binding energies, "[When] the nucleus is too big, repulsions start having a greater effect than the strong force."[4] In this case, testing students' planetary and cloud models of the atom using what if scenarios led to exposing an inconsistency between the models students believed in. This motivated a discussion of nuclear stability that was able to resolve the inconsistency. Students were able to use strong forces and binding energies later on to explain the nuclear stability of two new elements, helium and iron.

In summary, the structure of the what if scenario followed a general sequence of six teacher activities that included the teacher asking what if questions and the teacher promoting conceptual tasks in class. These six activities are labelled in Table 8.1.

Table 8.1 Structure of an educational what if scenario

Phase	Teacher activities
1	Teacher asks, "What if x is changed? How would it effect y?" (Where x and y are parameters of a model or conditions).
2	Teacher encourages students to run a test or a virtual test.
3	Teacher prompts students to generate a relationship between x and y.
4	Teacher asks another what if question and changes parameters in increments.
5	Teacher forms questions that interrogate the underlying model (here the teacher can ask what is wrong with the model, how does the model need to be changed; students can also inquire about hypothetical relationships in the model).
6	Teacher uses student inquiries to introduce new content or design further tests.

---

[4] Another point of discussion is that when the binding energy is not strong enough to hold the nucleus together, the atom is said to be unstable and this manifests itself as radioactivity. This presents an opportunity for the introduction of new information about atomic structure and nuclear stability.

8.5 Implications of What if Scenarios for Model-based Learning

The structure of a what if scenario is shown in Table 8.1. Returning to the same transcript in this chapter, students had identified several challenges to their original model of the atom after engaging in a what if scenario with their teacher. This is significant since students learned about the atom over several years of high school chemistry instruction. Understanding the relationship between electrostatic forces, charged particles, and distance was evident in the questions students asked. For example, one student appeared to explore the Coulombic relationship of distance and charge, and asked a question about the distance between the electron cloud and nucleus. Another student then recalled the planetary model, suggesting that the model of the atom included "rings". A student invoked the cloud model to ask what is between an electron cloud and the nucleus. Taken from the italicized portions of the transcript mentioned earlier in this chapter, the questions students asked such as:

> *S1 Why don't electrons pull into the protons?*
> *S2 Is the distance between the electron cloud and nucleus set?*
> *S7 Why don't electrons collapse into the nucleus?*
> *S9 Why do all the protons stick together in the nucleus?*

suggest that students' original models of an atom where electrons travel around a nucleus composed of protons were invoked. Students are calling into question how it is that the electrons don't collapse into the nucleus and why the protons stick together in the nucleus. A re-examination of a taken for granted planetary-like model of the atom is evident in this interaction where students were encouraged to test and revise their models. Provoking a re-examination of their atomic model is an important step towards improving the model.

8.6 Comparison of What if Scenarios with Traditional Teaching Methods and POE Tasks

This series of what if scenarios appeared to lay the groundwork for students to test and challenge their models of the atom in this episode. Students were able to test the charge and distance parameters on the electrostatic force between charged particles using a virtual "test bed", the Coulomb's Law Simulation. After testing, students constructed a working relationship between electrostatic force and distance between charged particles, and it appears that the teacher's question, "What's wrong with this picture" (referring to a drawn model of an atom), coupled with prior understanding of electrostatic forces, provoked student-generated questions about forces within the atom. Students were now in a position to enrich their models, having identified several questions that appear to highlight how electrostatic forces cannot account for subatomic particles and their interactions. The students had uncovered, according to their teacher, "a glaring problem" with their current model of the atom. The teacher now had an opportunity to respond to student-generated inquiries regarding their underlying conceptions of the atomic model and introduce new content.

The what if scenario method is different from telling students about electrostatic and strong forces at the outset, in that: (1) the information regarding electrostatic forces, distance and magnitude of charge has been gathered by the students from virtual tests, (2) the semi-quantitative and quantitative relationships among these variables have been co-constructed by students and their teacher, and (3) questions surrounding the atomic model have been generated by the students themselves. It is also different from traditional teaching in that students are first encouraged to recognize or regenerate their initial conception, then challenge that conception and revise it according to evidence and in a dialogic process with other students and the teacher.

What if scenarios can also be compared with Predict-Observe-Explain (POE) tasks common in science education (White & Gunstone, 1992). Both can be considered teaching strategies and student tasks designed to promote conceptual understanding of observable or unobservable phenomena. POE tasks probe student understanding by asking them to predict an outcome and justify their predictions, usually in an area where they are likely to generate a false prediction from a misconception. Students then make observations, usually of a discrepant event, that contradicts their prediction. They are then asked to explain the discrepancy in discussion in an effort to have them change their misconception. What if and POE share the basic strategy of producing dissonance with a student's prior conception or model in order to motivate the student to change the model. In a POE, there is a set sequence of activities that accomplish this; what if scenarios show that there are other sequences of activities that can do this, giving the instructor a number of options for promoting change. In what follows, I describe some of these options for instructors by describing places where the two techniques can differ.

The first difference is that the POE technique requires that the student make an initial commitment to a prediction. In what if scenarios, students can make a prediction based on the initial what if question; however, in this case, the teacher used the what if question to identify relevant variables and set the stage for running a test instead of asking students to make an initial commitment via a justified prediction. Both are valid approaches since either can lead to dissonance and modification.

In the second phase, POE encourages students to make observations via laboratory testing or teacher demonstration. An ideal demonstration, as explained by White and Gunstone (1992), is one that runs counter to common expectations. For example, in an ideal demonstration, students can predict what will have a higher temperature: boiling water or an equal volume of cooking oil that are both placed on the same hot plate. After students make their initial prediction, they are asked to observe the outcome of this test and describe what they see happening. It is a surprise to many students that the cooking oil will have a greater temperature. The second difference between POE and a what if scenario is that the pattern or model that will eventually participate in a dissonance relation in a what if scenario can be built up through a series of observations, rather than coming entirely from memory as a preconception in POE. That is, in a what if scenario, the new observations can come before activating the faulty model, whereas in a POE, the faulty model is activated before the new observations.

A third difference is that what if scenarios generate and test relationships explicitly, unlike POE tasks that ask students to explain their observations. In the present what if example, students worked with their teacher to generate a relationship between the variables identified in the first what if question; that is, the larger the charge, the stronger the electrostatic force, and the larger the distance, the weaker the electrostatic force. Testing can be run virtually as shown in this case, and this has the advantage that one can conduct more what if scenarios than laboratory testing might permit.

The fourth difference is that the teacher can employ the generated relationship to interrogate and probe the underlying model. This can be accomplished, for example, when the teacher asks questions to provoke a reexamination of a model and when students inquire about hypothetical relationships in taken-for granted models as they had done between electrons and protons in the atom in this case. In what if scenarios, the interrogation or probing period often produces more questions than answers. The probing period lays the groundwork for the teacher or students to introduce new content, postulate hidden factors, or design further tests. The teacher can play either a more or less leading role in introducing model modifications during the discussion after dissonance has been created. There are at least four differences between the POE method and the what if scenario, but it is likely that both what if scenarios and POE tasks can contribute to students' conceptual understanding via an enriched or modified model.

8.7 Conclusion

What if scenarios are useful teaching strategies to the extent that they support model-based learning. They are ideal to use when teachers would like students to interrogate and re-examine complex models in science, such as models of the atom, ecosystems, the weather, evolution, or human systems. What if scenarios can be used after a student has expressed an initial model of an observable or unobservable phenomena. It is at this point where teachers can support students' transformation of their original model with tests of the model. What if scenarios help students test parameters and explore the boundaries of their models. In the above case, the teacher asked students a series of what if questions that changed a single variable, such as magnitude of charge, in increments. Information communication technologies (ICT), such as java-based applets, microworlds, interactive animations, and hyperlinked tools, represent virtual "test-beds" to change variables associated with a model and observe the effects of the change. Here, what if scenarios appear to have led to an opportunity for students to express, test, criticize, reconcile, enrich and transform their models of the atom. What if scenarios extend the number of options teachers have for model-based teaching strategies and therefore should be useful in other areas of chemistry and in other subject domains.

References

Reiser, B. J., Tabak, I., Sandoval, W. A., Smith, B., Steinmuller, F., & Leone, T. J. (2001). BGuILE: Strategic and conceptual scaffolds for scientific inquiry in biology classrooms. In S. M., Carver, & D. Klahr (Eds.), *Cognition and instruction: Twenty five years of progress*. Mahwah New Jersey: Erlbaum.

White, R., Gunstone, R. (1992). Prediction-observation-explanation. In. R. White & R. Gunstone (Eds.), *Probing Understanding*. pp. (44–64). London: Falmer Press.

Chapter 9
Applying Modeling Theory to Curriculum Development: From Electric Circuits to Electromagnetic Fields

Melvin S. Steinberg

Smith College

9.1 Introduction

This chapter builds on Chapter 5, which described an introductory electricity course that is designed to foster model evolution using imagistic reasoning. The learning pathway in Chapter 5 modifies students' preconceived battery-centered model of movement in circuits into a model in which the agent of current propulsion is pressure in a compressible fluid in conducting matter. Curriculum design principles introduced in Chapter 5 are now applied to extending the learning pathway into the more complex realm of electrical distant action. Since moving electromagnetic fields require relativistic rather than classical mechanics for a complete quantitative description (Steinberg, 1983), this work pushes to the limit the idea of making highly abstract ideas make sense via concrete and schematic visualizable models.

New hands-on experiences reveal a need for additional model modifications that describe electrical distant action in terms of three increasingly abstract conceptions of causal agency:

(1) Charged objects influencing "pressure" in distant conductors motivates a conception of "pressure halos" around +/− charges analogous to temperature halos around flame/ice.

(2) Sparking in air between charged conductors motivates a conception of "electric fields" in pressure halos, with a local vector property that pushes +/− particles in opposite directions.

(3) Activation of a radio when current is turned on/off in a distant circuit motivates a conception of charge-pushing by dynamic "electromagnetic fields" created by accelerating charge.

The scarcity of imagery from concrete experience that transfers well to vector-field objects in space makes maintaining model runnability an important new issue. This chapter will:

- show experiments which demonstrate that electric and magnetic fields *contain* energy
- run field production simulations in which moving electromagnetic fields *carry* energy

I suggest that attributing properties as familiar as *containing* and *carrying* to fields may help students reify diagrams of vector fields in space as objects that can hold energy and exert force on charge.

9.2 Distant Action by "Pressure Halos"

9.2.1 Primary Discrepant Event

The circuit in Fig. 9.1 is designed to create a need to modify the compressible fluid model of current propulsion – in which charge is driven from a region of higher pressure to one of lower pressure – to include distant action. The model predicts that the bulbs in Fig. 9.1 will *not* light, since the insulating layers in the capacitors will prevent charge moving into or out of wires connected to the bulbs and thus prevent creation of a pressure difference in these wires.

But the bulbs *do* light (briefly). Why? A major clue is that a compass placed under the wire between the bulbs detects flow through the bulbs while the capacitors are charging, and then detects reverse flow when the battery is removed and the wires are reconnected to allow discharging.

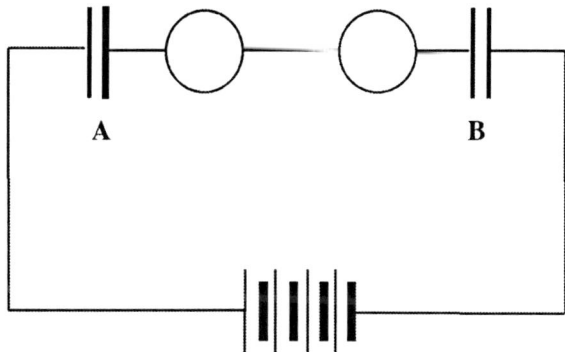

Fig. 9.1 Two capacitors prevent flow from battery to bulb

9.2.2 Framing the Problem

In the compressible fluid model developed so far, charge flow through each bulb is caused by a pressure difference across that bulb. Where is the compressed/depleted charge in which these pressure differences occur? Raising this question promotes class discussion based on the model – leading to the idea that the battery causes

- charge compression (+) and pressure above normal only in the left plate of capacitor A
- charge depletion (–) and pressure below normal only in the right plate of capacitor B

A useful teacher move would now be to draw (+) and (–) symbols by the extreme left and right capacitor plates in Fig. 9.1, and ask the class if there might be some way for these charges to change the pressure in the inner plates *across the non-conducting gaps* in the capacitors.

9.2.3 Revised Conception of Charge

Envisioning a mechanism of distant action is facilitated by a more complex conception of charge, which is developed in response to additional experiments that students perform at this juncture in the CASTLE curriculum. Space limitation requires that this conception be used here without presenting the evidence – from rubbed insulators and metal pie plates, with neon bulbs as directional flow detectors – on which the conception is based (see Steinberg et al., 2007):

- All normal matter contains *two kinds* of charge – pressure - lowering charge at (–) sites as well as pressure-raising charge at (+) sites. Neither kind is mobile in insulators.

- Normal matter has equal amounts of (+) and (–) charge, so that their effects cancel out.

The (+) and (–) symbols introduced in Chapter 5 will henceforth stand for an excess of one kind over the other – a terminological shift that occurred historically at the end of the 19th century. Bulb lighting in the circuit in Fig. 9.1 is consistent with a model in which

- (+) charge in capacitor A's left plate raises pressure across the gap in capacitor A

- (–) charge in capacitor B's right plate lowers pressure across the gap in capacitor B

9.2.4 Thermal Distant Action Analogy

Is there a good analogy for the mechanism of this electrical distant action? According to teachers piloting the CASTLE distant action models, an analogy that uses the imagery of distant action by flame and ice works well for beginning students: If (+) charge in the far left plate in Fig. 9.1 were replaced by a flame, the temperature in the right plate of capacitor A would be raised by a hot "halo" in the space around the flame. If (–) charge in the far right plate in Fig. 9.1 were replaced by ice, the temperature in the left plate of capacitor B would be lowered by a cold "halo" around the ice.

Most students can readily visualize hot and cold "halos" as containing a *condition in space* that makes one's finger hotter or colder without the finger actually touching the flame or ice to which the halos are anchored. Teachers using the CASTLE curriculum report that students like the terms "temperature-raising halo" around flame and "temperature-lowering halo" around ice. This language appears to help them transfer causal-agent "halo" imagery into the electrical domain as a "pressure-raising halo" around (+) charge in the left plate of capacitor A and a "pressure-lowering halo" around (–) charge in the right plate of capacitor B.

9.2.5 Halos Describing Electrical Distant Action

Teachers report a widely held student intuition that the effects of temperature-changing halos decrease with distance from flame or ice. Figures 9.2a and 9.2b are intended to help students transfer this intuition to pressure-changing halos around charges. They depict pressure-raising and pressure-lowering ability in the space around (+) and (–) charges, which diminish with distance from the charge. CASTLE calls this property "potential electric pressure", and calls the dashed curves "lines of equal potential pressure". This terminology differentiates the pressure-changing property from actual electric pressure, and identifies it as a closely related *potential to become* actual electric pressure in a conductor placed in a halo – e.g. in plates A and B in Figs. 9.2a and 9.2b. It also facilitates a later shift to the professional term "electric potential". CASTLE potential pressure halos are visualizations of mathematical electric potential functions used by professionals.

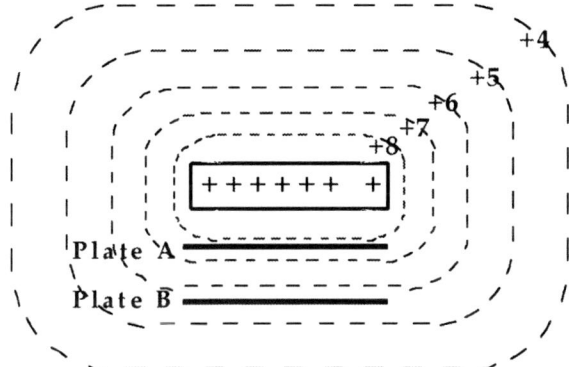

Fig. 9.2a Pressure- raising halo around (+) charge

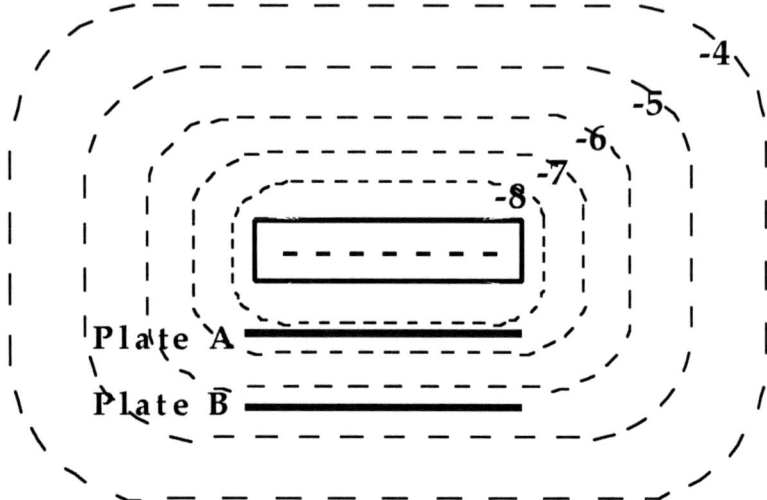

Fig. 9.2b Pressure-lowering halo around (–) charge

9.2.6 Investigating Halo Structure

An appropriate teacher move at this point in the curriculum would be to draw Figs. 9.2a and 9.2b *without* plates A and B – and then propose the following mechanism of halo action:

> *A conductor placed in a halo experiences potential pressure as actual pressure.*

Halo structures are then investigated by placing metal plates A and B in halos that are presumed to exist around sheets of plastic which have acquired (+) and (–) charges by being rubbed against each other. The halo in Fig. 9.2a predicts higher actual electric pressure in plate A than in plate B, while the halo in Fig. 9.2b predicts lower actual electric pressure in plate A than in plate B.

After placing plates A and B near a charged plastic sheet, students connect the plates to a neon bulb. A light flash from the bulb confirms that a potential pressure halo around the charged plastic has created an actual pressure difference in the metal plates. Knowledge from earlier experiments that a neon bulb lights only at its low pressure electrode enables students to use this observation to decide whether the halo has the structure described by Fig. 9.2a or Fig. 9.2b.

9.2.7 Application to the Circuit in Figure 9.1

The experimental confirmation of potential pressure halo structures around (+) and (−) charges unifies distant action by halos that bridge conducting gaps with local action by actual pressure in the mobile-charge fluid in conductors as complementary causal agents in circuits.

To help students predict events in the circuit in Fig. 9.1 using this unified model, they are asked to envision the charged rectangles in Figs. 9.2a and 9.2b as the extreme left and right plates of capacitors A and B – with (+) and (−) charges supplied by battery action. The three circuit diagrams in Fig. 9.3 depict a succession of predicted electric pressure values in wires connected to the bulbs, caused by pressure-raising and pressure-lowering halos around the charges in the extreme left and right plates. These figures use the color code for pressure values described in Chapter 4 to depict predicted bulb lighting events that students then actually observe.

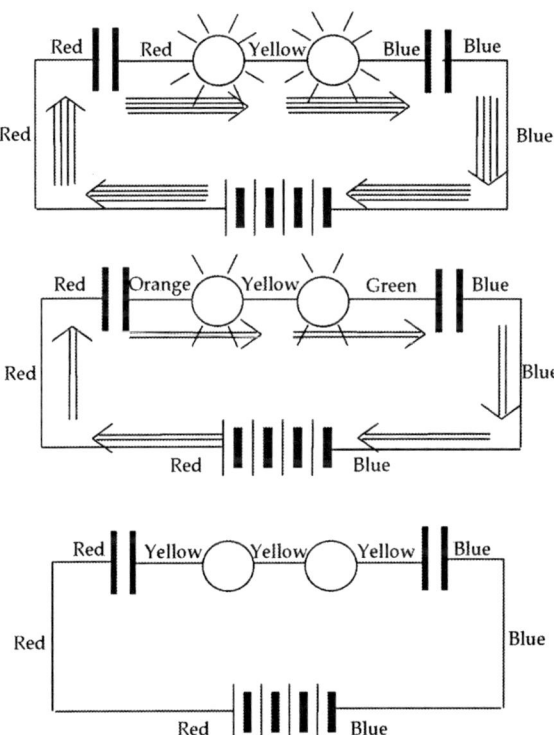

Fig. 9.3 Initial, intermediate and final charge flows in Fig. 9.1

9.2.8 Fostering Model Modification

The need to modify the compressible fluid model of current propulsion to include distant action across space that contains no matter was introduced by a discrepant event. A continued learning pathway was then developed, using additional principles of curriculum design that are listed at the beginning of Chapter 5. The salient design features of the new pathway are:

- Leading students through a sequence of steps that incrementally added the following three complexities of distant action to the compressible fluid model:
    o the existence in matter of pressure-lowering as well as pressure-raising charge
    o potential pressure in space around charge as well as actual pressure in conductors
    o HIGH/LOW potential pressure magnitude decreasing with distance from +/− charge

- Adding another analogy – temperature-raising and temperature-lowering "halos" around flame and ice – to make runnable imagery of pressure-changing halos available to combine with imagery of charge flow being propelled by pressure differences in electric circuits.

- Inventing diagrammatic conventions that provide external support for reasoning about halo action in analog and target domains. These have the simplest possible structure needed to make potential pressure values visible and convey other necessary information at this level.

9.3 Distant Action by Electric Fields

At the right of Fig. 9.4a is a capacitor made of aluminum pie plates separated by a foam cup. When this capacitor is charged using insulators electrified by rubbing, the extra plate at the left has the electric potential of the bottom capacitor plate to which it is connected. This plate is lifted by its foam-cup handle to a position above the top capacitor plate and then lowered until sparking occurs through the air between it and the top capacitor plate, as shown in Fig. 9.4b.

Fig. 9.4a Extra plate connected to bottom of pie-plate capacitor

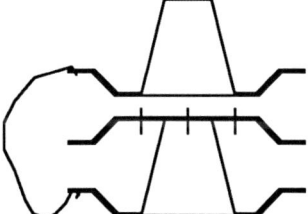

Fig. 9.4b Extra plate repositioned over top plate of the capacitor

Air is normally an insulator. But this experiment shows that atoms in air will break into (+) and (−) parts – making the air a conductor – if there is a potential difference across the air *and if the distance through the air is sufficiently small*. In Fig. 9.5a the upper and lower gaps between pie plates are too large. In Fig. 9.5b the upper gap has been made sufficiently small.

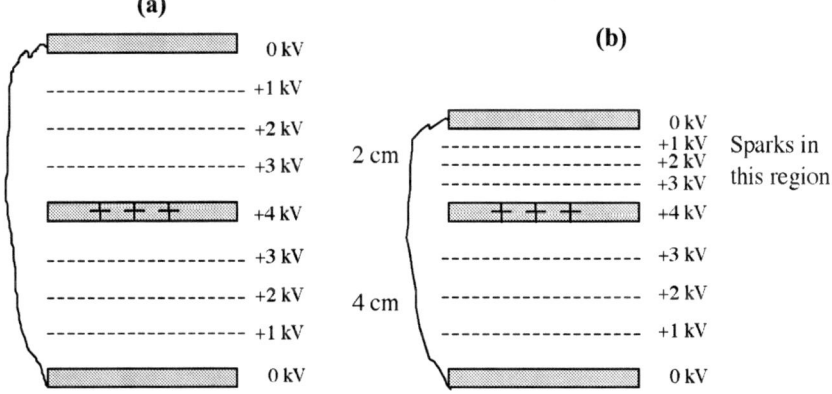

Fig. 9.5 (a) shows upper and lower air gaps are of equal width, while **(b)** shows extra plate is lowered to reduce upper air gap

9.3.1 Direct Action on Tiny Particles

Sparking between only the upper plates in Fig. 9.5b is evidence that atoms in air break apart when located where lines of equal potential are sufficiently crowded. Why is this *crowding* needed? A useful idea is that there is a previously unrecognized causal agent in the halo which:

- pushes the (+) and (−) parts of any atoms in a halo in opposite directions
- pushes more strongly where lines of equal potential are more crowded
- if strong enough, breaks atoms apart and makes an insulator conducting

This new charge-pushing agent, called "Electric Field", pushes directly on *individual* tiny charged particles. This is a more finely focused distant-action causal agent than pushing on a *collection* of charged particles by the potential difference across a macroscopic region of a halo. It is intended that an electric field be visualized as an *extended* object that possesses charge-pushing ability at every space point. This ability is represented by arrows drawn at various points, which show the direction the field will push on a tiny (+) particle placed at those points. Particles with (−) charge will be pushed in directions opposite to the arrows.

9.3.2 Electric Field Strength

The strength E of an electric field's charge-pushing ability at any point is related to the *degree of crowdedness* of equal potential lines in the vicinity of the point. The quantitative value of E may be determined by:

- choosing a space region of width Δx or Δy centered on the point
- noting the potential difference ΔV across this region
- quantifying degree of crowdedness as the ratio $\Delta V/\Delta x$ or $\Delta V/\Delta y$
- assigning a value $E = \Delta V/\Delta x$ or $E = \Delta V/\Delta y$ to the field strength at that point.

9.3.3 Representing Electric Fields Visually

The direction and strength of an electric field's charge-pushing ability at every space point are represented together by "electric vectors" that show electric field direction by arrow direction and pushing strength E by arrow width. Fig. 9.6 depicts the electric field of a charged capacitor. Numbers across the top are potential values for vertical equal potential lines that are not shown.

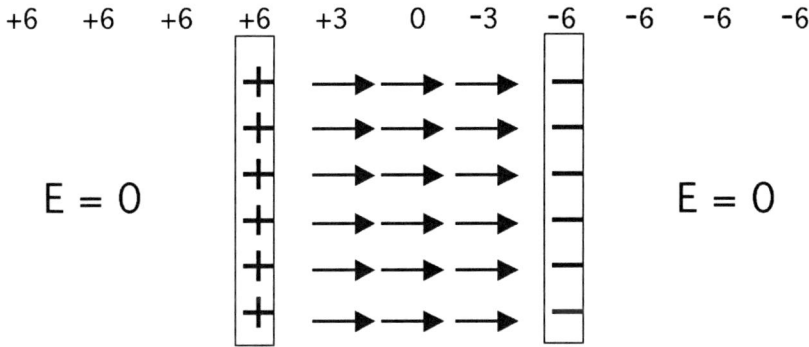

Fig. 9.6 Electric field vectors in space between capacitor plates

The diagram in Fig. 9.6 may be applied to the neon bulb used in investigating electric potential halos by regarding the capacitor plates as bulb electrodes and the charges as being placed on the electrodes by a battery. Though neon gas is normally an insulator, a sufficiently strong electric field between the electrodes will break the neon atoms into (+) and (−) parts, thus providing the mobile charge that makes the neon a conductor and allows the bulb to light. A sufficiently large potential difference across the bulb (typically 70 to 90 volts) introduces an electric field between the electrodes that is strong enough to make this happen. Neon bulb flashing differs from sparking in air only quantitatively – in the magnitude of electric field strength E needed to separate the (+) and (−) parts of atoms in the two gasses:

- in air between pie plates with fixed ΔV, make the separation Δy sufficiently small

- in neon between bulb electrodes separated by fixed Δy, make ΔV sufficiently

9.3.4 Energy Stored in an Electric Field

In Fig. 9.7a, a capacitor is charged by two D-cells and then discharged through a bulb. In Fig. 9.7b, the capacitor is charged by twice as many D-cells, placing twice as much +/− charge on its plates and giving it an internal electric field with field strength twice as great – and is then discharged through two bulbs. The fact that each bulb is brighter than the single bulb in Fig. 9.7a shows that a capacitor charged by twice as many cells stores more than twice the amount of energy. In Fig. 9.7c, the capacitor in Fig. 9.7b is discharged through four times as many bulbs as in Fig. 9.7a – with bulb brightness and length of lighting time the same as for the single bulb in Fig. 9.7a. This shows that a capacitor with E twice as large stores four times as much energy.

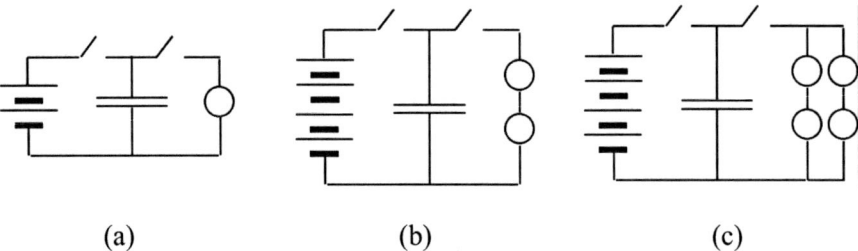

Fig. 9.7 (a) shows the capacitor charging and discharging through a bulb; **(b)** shows doubling the field strength and discharge through 2 bulbs; **(c)** doubles the field strength and discharge through 4 bulbs

When the capacitor is charged by three times as many cells to make E *three* times as large, the capacitor must be discharged through *nine* times as many bulbs in order for each to exhibit the same brightness and length of lighting time as the single bulb in Fig. 9.7a. The inference from this experiment is that energy stored in a capacitor is stored *in the electric field*, in an amount that is proportional to the *square* of the electric field strength in the capacitor.

9.4 Dynamical Electromagnetic Fields

While investigating the motor and generator effects, students use a compass to map the magnetic fields of magnets used in a motor. These vector fields extend the influence of North and South magnetic poles into the surrounding space, in much the same way that vector electric fields extend the influence of (+) and (–) electric charges into the surrounding space. Having encountered stationary electric and magnetic vector fields, students are ready to investigate the more complex domain of dynamic electromagnetic vector fields. They connect and then disconnect a battery in a circuit with a wire – and listen to a *radio* located a meter or more away. The radio is tuned to a quiet place between active stations at the low frequency end of the AM band. It makes a momentary "popping" sound when the battery-&-wire circuit is closed – and again when the circuit is opened. It remains silent while the circuit remains closed.

9.4.1 An Electric Field that's Moving?

The sound from the radio is evidence that charge in a circuit in the radio is being made to move. This surprising phenomenon raises three successive questions.

1. *What causes this charge to move?* Electric field is the only causal agent students are aware of that can make the tiny charge carriers in a circuit move. Moreover, electric fields have been shown to contain energy – which could be transferred to the radio and enable it to produce these sounds. So the most plausible hypothesis is: *An electric field causes the movement.*

2. *How does the field get to the radio?* Since this electric field is not normally present in the radio, it must move into the radio from elsewhere. There is strong circumstantial plausibility that the electric field acting on charge in the radio is *created in* the battery-&-wire circuit when charge speeds up or slows down, and *moves out* of the circuit in all directions and reaches the radio, which it then acts on

3. *How is the moving field created?* Students will note that sounds from the radio occur only when charge flow in the battery-&-wire circuit is turned on or turned off – and they will be able to associate creation of moving electric field with charge speeding up and slowing down.

9.4.2 Investigation of Field Creation and Movement

The wire in the battery-&-wire circuit is now replaced with a tubular wire coil, as shown in Fig. 9.8. This is done to provide spatial clarity by dividing space into two parts:

- Exterior space – in which an electric fields will move away from the wire where it is created

- Interior space – where radiated electric fields will cancel out and a stationary magnetic field will form from radiated magnetic fields

The LED can detect electric field *created in the coil wire*. It will not glow if the wire is only a zero-resistance flow path that short-circuits the battery. But it will glow just after switch closure

- if the speeding-up of charge that occurs then *creates electric field* in the wire, and

- if the created electric field *opposes charge flow* so it *acts like resistance* in the wire.

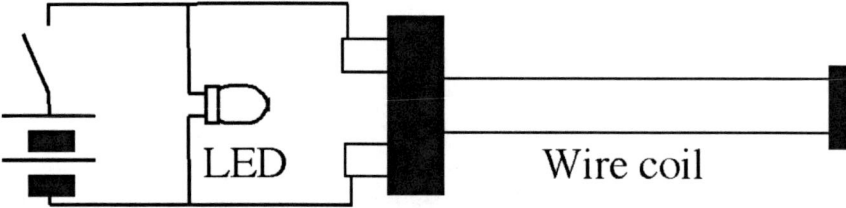

Fig. 9.8 LED shows accelerating charge creates electric field in wire

Students can see that the LED glows – indicating that an electric field is created in the wire, with a direction that opposes charge flow. And they can see that it glows *briefly* – indicating that the field moves quickly out of the wire it is created in. These observations support the idea that the "popping" sound they hear from the radio is caused by

- charge in a radio circuit being made to move by an electric field, which is
- created in the coil wire by accelerating charge when the switch closes, and
- moves out of the coil wire in all directions, including that toward the radio

9.4.3 Simulating the Field Radiation Process

The electric field being created in and moving outward from the coil in Fig. 9.8 after the switch is closed, when mobile charge in the coil is accelerating, is illustrated as a graphic simulation in Figs. 9.9a and 9b. The arrows marked **E** show electric field directions opposite to arrows marked **I** that show circuital charge flow in the coil wire while charge in the wire is speeding up. Unmarked arrows pointing to left and right show movement of electric field out of the wire. The direction of movement is:

- into the interior space – where counter-moving electric fields overlap and cancel out
- into the exterior space – where parallel-moving fields cancel partially overlap and cancel

This cancellation of oppositely directed quantities appears intuitive to teachers and students.

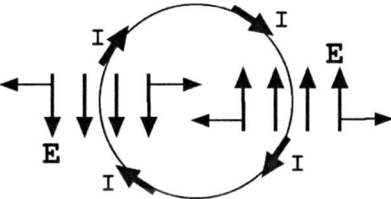

Fig. 9.9a Electric field opposing charge flow while moving out of wire

After fields moving into the interior space cross to the other side of the coil and cancel out where they overlap, there remain *radiated pulses* of electric field the width of the coil moving away from the coil in both outward directions.

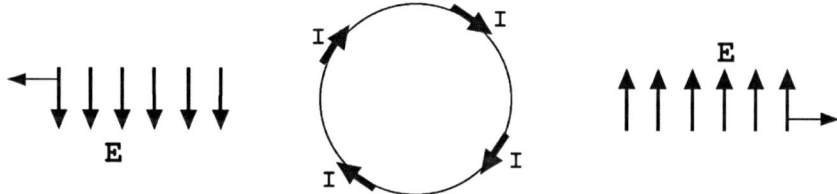

Fig. 9.9b Radiated electric field moving away from coil to left and right

9.4.4 Need for a Companion Magnetic Field

It is important to focus students' attention on moving electric field *before* recognizing the presence of a companion magnetic field in the space around wires with turned-on current. If the electric field is investigated after the magnetic field, it can seem plausible that the electric field detected by the radio is "caused by the changing magnetic field" when the current is turned on and turned off. This explanation, which also appears in many textbooks, confuses correlation with causality. (The fields occur together, but one does not produce the other.) Its plausibility disappears in workshops where teachers participate in building and running a dynamic model of magnetic field production that first investigates the moving electric field.

Investigation with a magnetic compass will detect a stationary magnetic field in the space around the wire *while there is steady charge flow* in the wire – during the time interval between flow turn-on and turn-off. It is clear that this magnetic field *contains energy*, since it is the only possible source of the energy for sound production by the radio *after the battery is disconnected*.

The stationary magnetic field around current-carrying wires must acquire its energy in the process that creates the moving electric field. A major question of field production is then: What is the mechanism by which these energy-containing fields are created *together*? This question is not raised in conventional curricula, which describe the magnetic field simply as being "turned on". It has become apparent in CASTLE workshops for teachers, however, that developing a dynamic imagistic model that can be run to describe *how stationary magnetic fields are created* is key to qualitative understanding of how vector electromagnetic fields are produced.

The process of creating the complete electromagnetic field is described in Figs. 9.10a and 9.10b. The magnetic vector directions in these diagrams are designated by the symbols (•) for tip of arrow and (**X**) for tail of arrow. The directions are chosen to relate the stationary magnetic field that is created in the interior of the coil to the direction of charge flow in the coil wire according to the professional "right hand rule" (Steinberg, 1984).

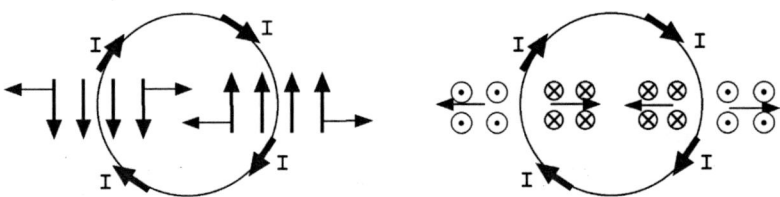

Fig. 9.10a Right-hand-rule magnetic field moving with electric field

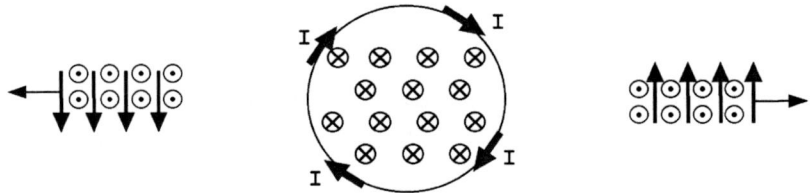

Fig. 9.10b Magnetic field and electromagnetic radiation created together

9.4.5 How is a Moving Electric Field Created?

Consider Fig. 9.11a, which is what Fig. 9.6 would look like if the (–) charge in normal matter were not present. Regard the vectors in Fig. 9.11a as depicting only one feature of an *object-like* electric field in the space near a large charged (+) plate. Now imagine *another feature* – depicted in Fig. 9.11b as "field lines" drawn head-to-tail through the vectors. The following additional characteristics of the object-like electric field can be identified:

- Field lines terminate only at source charge. They are continuous throughout empty space.

- Field lines exist only when source charge is present. They appear *attached* to the charge.

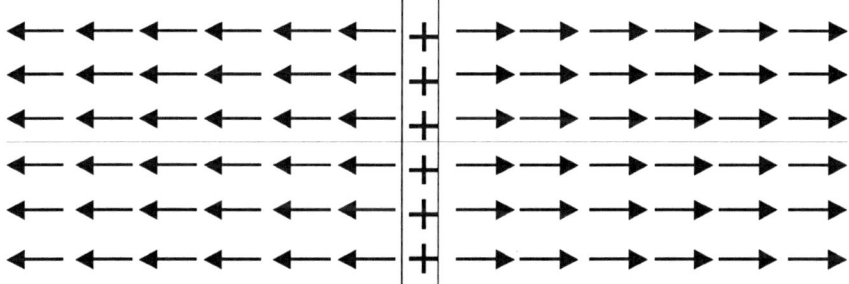

Fig. 9.11a Electric field vectors in space near mobile charge in a wire

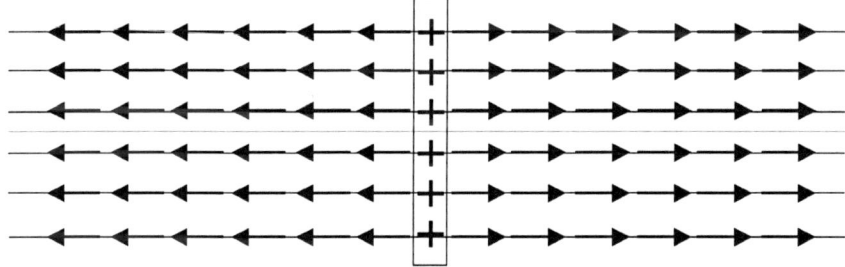

Fig. 9.11b Including field lines to which electric vectors are parallel

Now imagine the charge in Figure 9.11b to be jerked upward – and the attached ends of the field lines jerked upward with it. These lines cannot break: That would create new line ends, which would require creating new charge. Field line integrity can be preserved if the field lines are *elastic* – and form *kinks that move outward* from the charge:

- These kinks are created where the charge is accelerating.
- They move outward from the place where they originate.

As illustrated in Fig. 9.11c, the electric vectors in the kinks point downward, transverse to their original direction and opposite to the charge's direction of movement. The elastic field lines in this simulation appear to *behave like "slinkies", which give the vectors a holistic dynamical organization* – an imageable concrete analogy.

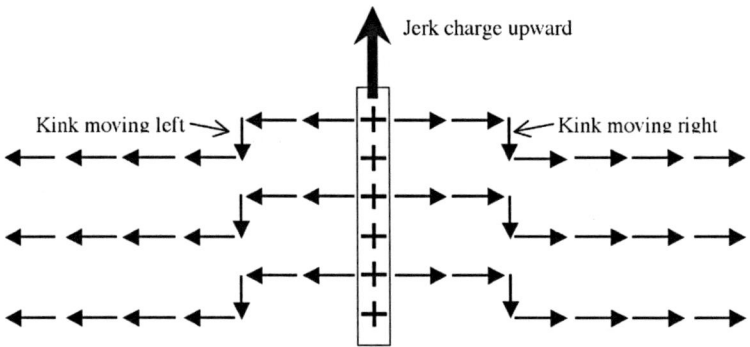

Fig. 9.11c Accelerating charge causes moving kinks in elastic field lines

The outward-moving kinks in Fig. 9.11c match the structure and movement of electric field in the field production simulation depicted in Figs. 9.9a and 9.9b. Thus, it makes sense to regard moving electric field created when charge is accelerating as *a propagating disturbance in an electric field* that is stationary in the space around the charge prior to acceleration.

9.5 Maintaining Model Runnability

The most salient curriculum design features carried forward from Chapter 5 have been those that added incremental complexities to the "halo" model of electrical distant action:

- vector electric fields push directly on tiny charged particles
- particles with (+) and (−) are pushed in opposite directions

The added conventions that make potential pressure levels visible and provide external support for reasoning about vector field action have the simplest possible structure needed to convey the information necessary at this stage of model building.

But can model runnability be maintained while introducing increasingly abstract causal agents to account for increasingly complex phenomena? Classroom trials and workshops for teachers have stimulated keen participant interest in the discovery that the amount of energy stored in a charged capacitor is determined by the strength of the electric field in the capacitor. Participants' have also shown great interest in experimenting with and reasoning about moving vector fields. One may speculate that this interest is aroused by the perception that electric fields are *containing* and *carrying* energy – thus providing evidence for their being object-like, which is a basis for imageability. This may help students reify electric field diagrams, and could enable them to think of vector field diagrams as being like images of concrete objects that can push charges in determinate directions. It is an interesting question for formative research.

There are two additional possible resources, which are as yet untested in the classroom, for maintaining runability of models of dynamical electromagnetic fields:

1. transferring imagery from observations of slinkies into the domain of dynamic fields
2. inferring energy of moving electromagnetic fields from progressive interpenetration

Resource (2) is not discussed here, due to lack of space for dynamic interpenetration diagrams.

9.6 Status of Materials Development

CASTLE materials on electromagnetic fields and devices are presently in a late stage of development. They are intended to foster qualitative understanding of electromagnetic field production and action – which is poorly developed by conventional instruction (Bagno & Eylon, 1997; Raduta, 2003). The materials are being tested in several high school classes and in workshops for teachers at meetings of the American Association of Physics Teachers. Both have provided feedback that has helped the development team improve instructional materials. The workshop participants have provided good written and oral evaluations, which include comments about having begun to understand electromagnetic fields for the first time.

9.7 Conclusion

My concluding claim is that the curriculum ideas discussed here, which apply to electrical distant action, embody many of the design principles described at the beginning of Chapter 5 on electric circuits. In particular, the progression of diagrams in Figs. 9.2 to 9.6 represents a careful building up of a qualitative model that is designed to enable students to make sense of the vector electric field concept. This progression embodies principle **P-4** in Chapter 5 – Gradual Model Modification – whereby sense-making continuity and model runnability are maintained via small-step model revisions.

This series of diagrams can also be seen as an extended development according to principle **P-8** – Imagery Enhancement – in which external drawings add important detail to imagery. In addition, Fig. 9.6 serves as a contrast diagram that makes visible the distinction between scalar electric potential and vector electric field – two concepts and representations that could easily be confounded. At the same time, it shows the connection between the concepts. Electric fields are completely invisible, and these diagrammatic tools are designed to make them imageable.

The progression of diagrams in Figs. 9.9a to 9.10b applies the same principles to running a simulation to evaluate a model modification that adds *moving electromagnetic* vector fields. The modification of Fig. 9.11a to include field lines that link vectors in Fig. 9.11b, with elasticity that enables them to propagate disturbances like "slinkies" in Fig. 9.11c, illustrates the use of multiple analogies in connection with principle **P-6** – Transfer of Dynamic Imagery (Clement & Steinberg, 2002). Here, concrete dynamic imagery from an additional analog domain is transferred into the target domain – in this case helping to make the construct more convincingly complete and runnable.* Thus as we have become more articulate about and aware of learning principles in modeling theory it has had an increased impact on our curriculum development efforts in electricity.

---

*The learning path followed here shows similarities to that followed by James Clerk Maxwell in his early electromagnetic field modeling (Maxwell, 1864) which utilized multiple analogies – vortices, followed by gears and idler wheels – to provide imageable constraints as the model evolved to become more complete and productive. Nersessian (2002) points out that Maxwell used analogical elements from both fluid mechanics and machine mechanics.

References

Bagno, E., & Eylon, B-S. (1997). From problem solving to a knowledge structure: An example from the domain of electromagnetism. *American Journal of Physics, 65*, 726–736.

Clement, J., & Steinberg M. S. (2002). Step-wise evolution of models of electric circuits: a "learning-aloud" case study. *Journal of Learning Sciences, 11*(4):380–452.

Maxwell, J. C. (1864). A dynamical theory of the electromagnetic field. *Proceedings of the Royal Society, 13*, 531–536.

Nersessian, N. J., (2002). Maxwell and "the method of physical analogy": model-based reasoning, generic abstraction, and conceptual change. In D. B. Malament (Ed.), *Reading natural philosophy*, pp. 129–166. Chicago, Open Court.

Raduta, C. (2003). *General students' misconceptions related to electricity and magnetism*. Doctoral dissertation, Ohio State University.

Steinberg, M. S. (1984). Electromagnetic waves without Maxwell's equations. *American Journal of Physics, 51*, 1081–1086.

Chapter 10
Developing Complex Mental Models in Biology Through Model Evolution

M.C. Núñez-Oveido

Universidad de Conception

John Clement

University of Massachusetts, Amherst

Mary Anne Rea-Ramirez

Western Governors University

10.1 Introduction

This chapter presents new vocabulary and conceptualizations for describing teaching strategies that foster teacher-student co-construction of mental models in large group discussions of complex topics. We will focus in particular on an overall teaching strategy called Model Evolution that emerged from detailed analyses of videotaped lessons and protocols with middle school students in the area of human respiration. Model Evolution is a teacher-student interaction process through which students restructure their initial ideas to produce successive intermediate models, until, hopefully, reaching the target model for the lesson. The Evolution process utilizes several sub-processes, such as Element Disconfirmation, Element Confirmation, and Model Modification, and we present diagrammatic models of their contributions. These vocabulary terms and conceptualizations will be used to examine two examples of large group discussion in which concepts of circulation and diffusion are built. Our long-term goal is to make these implicit modes of teaching explicit, so that teachers are more aware of them. This will provide teachers with new and important strategies for fostering model construction.

In conjunction with the other authors, Nunez-Oviedo (2003) has developed vocabulary and diagrams to explain the processes through which a middle school science teacher supported her students in building mental models in the area of human respiration. Accounting for all of the major cognitive strategies identified in the transcripts pushed us to define several basic conceptual change processes and teacher-student interaction modes. In Chapter 7, three teacher-student interaction patterns were discussed that are complementary to those described in this chapter. We will now describe another essential co-construction mode, called the Evolution Mode, that was used by the same teacher. In addition, we will describe four smaller processes that can be involved within the Model Evolution Mode: a Disconfirmation Mode, Confirmation Mode, Modification Mode, and Accretion Mode. We will first define these modes in a theoretical way and then combine them in explaining two classroom episodes.

Some of the previous work that analyzes group discussions in classrooms has focused on student argument structures (e.g. Osborne, et al., 2001; Kelly, et al., 1998; Duschl & Osborne, 2002). Such analyses generally have been based on schemes developed by Toulmin (1972) for the analysis of hypothesis evaluation structures in science. The study reported in this chapter uses an alternative approach. Instead of focusing only on student arguments, we also focus on model-based strategies that the teacher uses to interact with student thinking. In addition, instead of using the pre-established models of argumentation and reasoning of Toulmin, we begin from naturalistic observations of classroom interactions, and then attempt to build an explanatory model of what is occurring. To build the explanatory model, we used a method for generative exploratory studies described in Clement (2000), derived from the constant comparative method of Glaser and Strauss (1967). This way of modeling student interactions is also related to a previous study conducted in the area of classroom dialogue (Resnick, Salmon, Seitz, Wathen, & Holowchak, 1993).

10.2 Teacher-student Interaction Patterns

In Disconfirmation Mode, a model is criticized by students or the teacher, and, if this criticism is strong enough, the model drops out of the classroom discussions (Fig. 10.1). The disconfirmation process is the result of fostering an episode of dissonance via a tactic such as a discrepant event, discrepant question, or thought experiment (Nunez-Oviedo, Rea-Ramirez,

Clement & Else, 2002). In Fig. 10.1, time moves from left to right, and the dissonance-producing event introduced by the teacher or a student (T/S) causes Model M1 to be disconfirmed. The dissonance is strong enough that the model is discarded, symbolized by the X. In using this term, we do not necessarily imply that the idea has completely disappeared from the students' minds. However, when such a model does not reemerge in the discussions, there is some evidence that the students considered it to be disconfirmed.

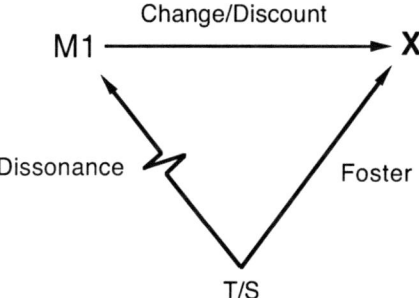

Fig. 10.1 Disconfirmation mode

In Modification Mode (Fig. 10.2), a similar process takes place, except that, instead of discarding the model, the students modify model M1 to repair or extend it. This can be successful when M1 is at least partially compatible with the target concept. Figure 10.2 shows model M1 being questioned by the teacher or a student (T/S), causing dissatisfaction, and then modified to concept M2.[1] The arrow from T/S to M2 indicates that such an action is expected to constrain the modification.

---

[1] Unlike Fig. 10.1, Fig. 10.2 shows a "dissatisfaction" relation instead of "dissonance" to indicate a closely related but broader idea. Building on Hatano and Inagaki (1986), Rea-Ramirez and Clement (1999) defined *dissonance* as a sensed internal discrepancy between a conception and another entity (observation or other conception). *Dissatisfaction*, on the other hand, refers to a broader category, which includes dissonance but can alternatively include a state that is not as strong: a sensed incompleteness of an explanation provided by the model. This reflects the case where there is nothing directly conflicting with the model, but the model does not provide an explanation for some particular aspect of the target.

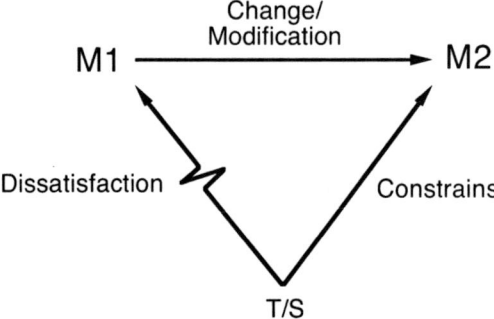

Fig. 10.2 Modification mode

We can identify three subtypes within the Modification Mode: (1) Adding a Model Element; (2) Removing an Element; or (3) Replacing an Element.

The Evolution Mode consists of two or more consecutive Modification Modes (Fig. 10.3). The process may need to be initiated by Model Generation of M1. The teacher uses the Evolution Mode to produce successive intermediate models labeled as M2 and M3. These evolving changes occur as a result of teacher and student interactions that produce the addition, the modification, or the removal of model elements over time. The jagged lines indicate dissonance or dissatisfaction with a prior model caused by an event such as a discrepant question, analogy, or a thought experiment. The diagram also shows how the students may contribute to the process from their prior knowledge of useful ideas (schemas), labeled S.

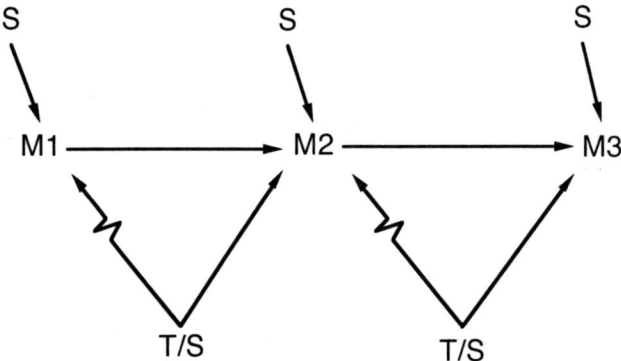

Fig. 10.3 Evolution mode

The Model Generation Mode (Fig. 10.4a) is often triggered by the teacher making a request for an explanation of a target phenomenon.

Assuming students have not already been given an explanation, they must invent an explanatory model that explains why the phenomenon occurred. Sometimes a student will volunteer an explanation before the teacher requests it.

In the Confirmation mode, the teacher and the students provide evidence to support an initial hypothesis (Fig. 10.4b). Because of this mode, the initial idea may pass or survive the teacher's or the students' evaluation. The Confirmation and Disconfirmation Modes can be applied either to an entire model (as was primarily the case in Chapter 7) or to a model element.

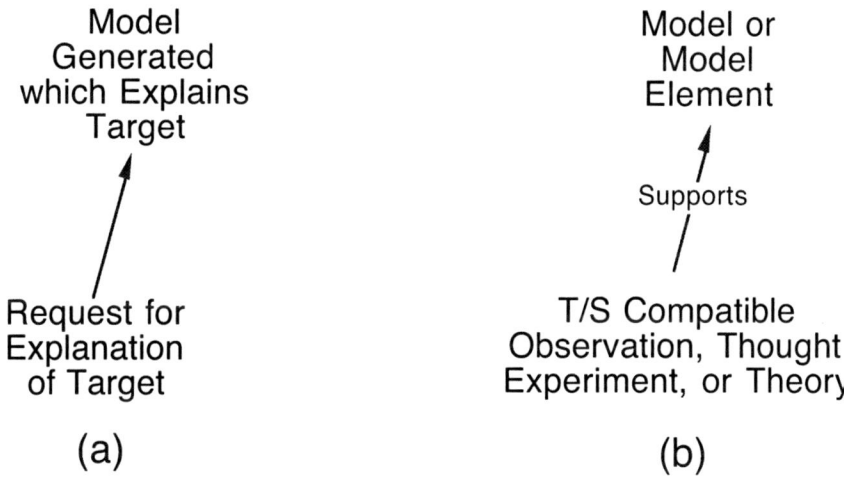

Fig. 10.4 (a) Model generation mode; **(b)** Confirmation mode

The above modes can be thought of as basic processes for learning or developing scientific models. On the other hand, what we call the Accretion Mode is a more specialized type of teacher student-interaction pattern that is common in classrooms. In the Accretion Mode (Fig. 10.5), there is a repeated teacher-student interaction pattern that consists of a teacher's question (Q), student's answer (Sn), and the teacher's positive feedback (PF). The teacher provides the students with positive feedback by using words such as "good", "okay" or by immediately asking the next question. As a consequence of this pattern (Q-Sn-PF), there is a selective accretion of elements of the model (+abc) that are compatible with the target of the lesson. By asking a series of leading questions about easy ideas that students can infer, the teacher helps the students put together a string of small model elements where every piece contributed is essentially correct.

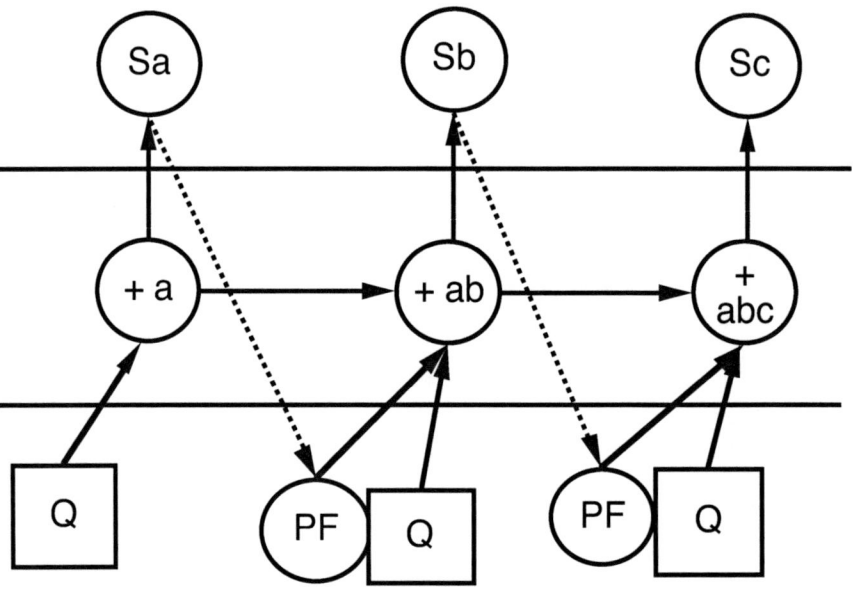

Fig. 10.5 Accretion mode

10.3 Combining Co-construction Modes

In what follows, we consider two instructional sequences exemplifying the Evolution Mode. These utilize the smaller modes above as cognitive subprocesses. We will combine the diagramming elements above to illustrate the overall learning pathways that took place during the construction, revision, and modification of successive intermediate mental models of human respiration (Figs. 10.6, 10.7, 10.8 and 10.9).

The upper level in these Figures shows students' statements as contributions to the co-construction process. The center row shows the students' successive intermediate mental models (M1, M2, etc.) leading to the target concept. A horizontal line stretching to the right of an element means that the element is still actively under consideration. The bottom layer of dialogue shows the teacher's statements as contributions to the process. Below this are shown the interaction modes that describe our hypothesis about the teaching strategies being used. In this view, the Disconfirmation Mode, the Modification Mode, the Confirmation Mode, and the Accretion Mode, are basic processes that lead to the evolution of the students' ideas illustrated in the center row and reflect implicit teaching strategies being used to promote conceptual change. These strategies therefore, are classified in terms of their intended effects on the students' model construction process.

The description in the middle rows of the diagram denotes the elements of the current model. Of course, individual student models will vary; some students will have a more sophisticated and some students a less sophisticated model than the one shown. In generating these descriptions, it was hypothesized that they describe the teacher's view of what an "average" student's mental model is at that point. The intention is to represent the hypothesized student model that the teacher is using to base his or her next "move". We based these descriptions on the student and teacher statements, shown at the top and bottom of the diagram, as well as their drawings.

By using the diagrams that illustrate each individual mode and combining them to represent "learning pathways" (Scott, 1992; Rea-Ramirez, 1998) (Figs. 10.6, 10.7, 10.8 and 10.9) we attempt to show an explicit description of implicit teaching strategies used and the hypothesized effects that they can have on student mental model construction processes. This extends the diagramming and analysis schemes developed by Nunez-Oviedo MC (2003), Rea-Ramirez (1998, 2005), and Clement and Steinberg (2002).

10.3.1 Example 1

The first example was recorded in a seventh grade class during a unit on circulation. The teacher was engaged in a co-construction process with the students to develop an understanding of the presence of "valves inside of the veins" to help the return of venous blood flow back to the heart. The scientific model states that the continuous pump of the heart is not sufficient to push the blood flow in the legs uphill against gravity. The process is aided by calf muscle contractions that push blood upward. Between these contractions, one-directional valves close off to prevent blood from backwashing. Figure 10.6 shows a "learning pathway" for developing the valves idea inside veins via model evolution. We will discuss first the sub-processes involved and then the overall Evolution Mode.

At the beginning of the episode, the teacher used an analogy in which she compared the pumping action of the calf leg muscles to the act of squeezing a toothpaste tube (Fig. 10.6) to explain the upward movement of blood in the veins. However, there was a problem with this analogy since in reality, the muscle must contract and relax and the blood could flow back down when the muscle relaxes. Toothpaste, on the other hand, might not have this problem since it would not flow back down in a tube because it is "pasty". The teacher, therefore, asked what would keep the blood from flowing back down in the veins. A student suggested that the blood is pushed along by the blood behind it. There is some pressure in the veins, so this element is a modifying addition to the model that is partially correct, but it turns out not to be the full story. Blood pressure is decreased considerably in the capillary pool and, therefore, there is not sufficient force from below on return through the venous system up to the heart.

Therefore, the teacher drew out this disanalogous aspect of the toothpaste tube analogy by asking, "Is blood a solid pasty liquid?" Several students answered "no" and she offered positive feedback by repeating the students' response. Our hypothesis is that the teacher believed this would cause dissonance with part of the model that had been developed at that point because it removed the only plausible mechanism up to that point that could explain why the blood could not flow back down away from the heart when leg muscles are relaxed. This is shown by the arrow labeled "dissonance" in Fig. 10.6.

The teacher then helped the class to generate the idea of the presence of valves inside the veins and how they work to explain the upward movement of the blood. The teacher suggested to the students that within the veins there was a "design" that prevented the blood from going back down. A student proposed that perhaps within the veins there were some kind of "fingers" or "brush-like" structures. Even though these ideas are somewhat distant from the target concept, they are very valuable because they constitute a good starting point to build up the concept of "valves". Thus, the toothpaste element was replaced by the finger structure element to keep blood from flowing back down. This is the second modification in a series of model modifications shown at the bottom of Fig. 10.6.

The teacher then introduced a second analogy to replace the "finger" or "brush" like structure idea inside the veins. To do this, the teacher made a connection to prior student learning by asked the students to describe what structures help to prevent choking. Students indicated that a "flap" closes over the airways. The teacher offered positive feedback and suggested that veins might have something similar inside. As a result, a student generated a spontaneous analogy to explain the upward movement of the blood. She suggested that structures might work like a "lobster trap" in preventing the reverse flow of blood inside the veins. The teacher provided positive feedback and helped the rest of the class in understanding the functioning of a lobster trap. She also introduced the term "valves" to replace the term flaps to name the structures that are inside of the veins that work like a "lobster-trap" in preventing backflow. At this point, she appeared to believe that the

Developing Complex Mental Models in Biology

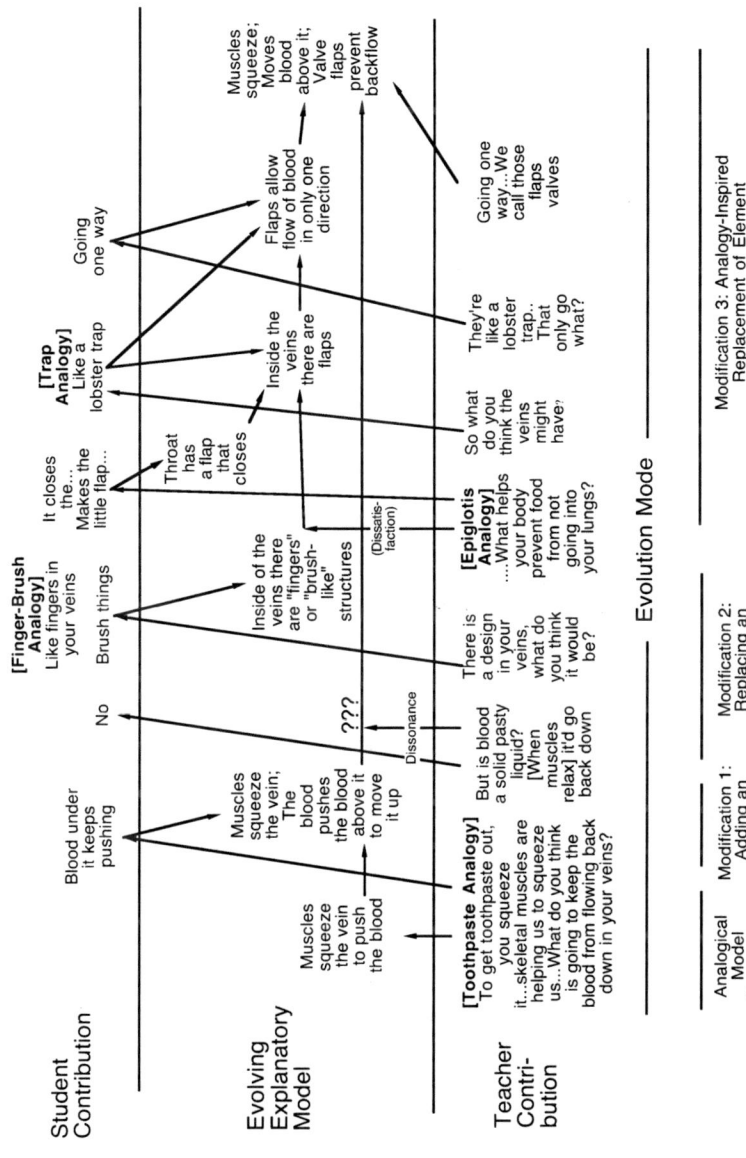

Fig. 10.6 Learning pathway showing the evolution of the concept "valves inside the veins"

students had reached the target concept in understanding that the valves of the veins are like a flap that allow blood only moving in the upward direction. Based on the protocol, the teacher did not simply give the students the information about valves inside of the veins. Instead, she, and the students, exchanged ideas back and forth about the topic, co-constructing the concept step by step.

At the point where the students propose the "brush-like" structures in the veins and the teacher responds with the "flap in the throat" (epiglottis) analogy, Fig. 10.6 shows the hypothesis that this analogy creates dissatisfaction with the existing "fingers" model. This reflects the view that when the teacher introduces an analogy or contributes an idea to a model, he or she may introduce "extrinsic dissatisfaction" with the existing model by implying, in effect: "If you think about this additional factor, you may gain more understanding of it than you have already." If the students want to understand the target, this signals to them that they should be dissatisfied with their present model because something more is needed according to the teacher. These relations are shown in the diagram as an arrow labeled with "(dissatisfaction)" in parentheses. Extrinsic dissatisfaction may be less desirable than intrinsic dissatisfaction, but it is a common occurrence in tutoring and teaching.

In summary, after initiating model generation via an analogy, the teacher promoted an extended episode of Model Evolution (Fig. 10.6). She did this by developing a series of three subsequent mental models of the functioning of the valves inside the veins that became progressively more adequate: (1) blood moves up because "the blood is pushed along by the blood coming behind it"; (2) there is a "design" inside of the veins that look like "fingers" or "brush-like" structures; (3) there are "flaps" called "valves" inside of the veins that prevent backflow. This involved using two kinds of model modifications: adding an element, and replacing an element. This episode recalls the distinction made in Chapter 1 between a *teacher directed* (as opposed to student directed) lesson and a *teacher generated* lesson. Although the discussion is rather tightly directed by the teacher in this case, she succeeded in having students generate many of the ideas in the dialogue. Thus, the dialogue is largely teacher-directed as opposed to student-directed. However, it is also largely student-generated as opposed to teacher-generated, in the sense that students generate many pieces of the ideas. The teacher helped by generating two analogies, but also succeeded in stimulating the students to generate two creative model modifications, and to infer a third one, with two creative, two student generated analogies along the way. We refer to this as teacher-student co-construction. The teacher supported the students in evaluating and modifying (evolving) their mental model until arriving at the targeted mechanism to explain the upward movement of blood in the legs. This interaction is different from "recitation" because it is not going over material that the students have learned previously.

10.3.2 Example Two

A second episode lasted almost forty minutes and was recorded in a seventh grade class (G4) during the teaching of the circulation unit. The topic of the lesson was the concept of diffusion as the process of the movement of particles from an area of high concentration to an area of low concentration. The concept of diffusion is essential for understanding pulmonary respiration, digestion, and exchange of chemicals between the blood and the cells through the capillaries, in this curriculum. To teach the concept of diffusion, the Energy in the Human Body Curriculum (Rea-Ramirez, Nunez-Oviedo, Clement, & Else, 2004) suggests conducting a demonstration activity that includes two steps.

In the first step, the teacher placed a raw egg into a vinegar solution. The students, who were asked to conduct observations over a five-day period to complete the entire activity, found that the eggshell dissolved and that the shell-less egg enlarged. (This is caused by vinegar diffusing into the egg.) In the second step, the teacher placed the enlarged egg on top of green colored corn syrup in a jar. The students then observed that, after some time, the egg shrank, allowing them to see the yolk through the membrane; that a thin liquid was present at the top of the jar with a thick liquid underneath; and that they could smell vinegar solution in the jar. (The vinegar diffuses out of the egg and floats on the top of the corn syrup.) Based on classroom videotape data, Figs. 10.7, 10.8 and 10.9 show our hypotheses concerning the learning pathway used by students to reason about this experiment, and how the concept of diffusion evolved.[2]

The episode began when the teacher asked the students to discuss within their small groups why some liquid had come out of the egg. After about 5–10 minutes, the teacher asked the students to share their ideas with the large group and asked one group (Group 5) to provide an explanation. The students generated the idea that the vinegar went out of the egg and that this was because the corn syrup went inside, pushing out the vinegar (Fig. 10.7). The first half of this idea is correct, but the second half is not compatible with the scientific explanation.

---

[2] Since this includes details of actual student reasoning, this is a much more detailed learning pathway than those described in other chapters as part of a planned curriculum. When we need to distinguish between these two pathways, we can refer to a "planned learning pathway" and an "implemented learning pathway".

To deal with this, the teacher, without indicating whether Group 5's idea was incorrect, encouraged the rest of the groups to react. The students at Table 10.2, appeared to agree that the vinegar came out, but challenged Group 5's explanation that the corn syrup entered the egg and displaced the vinegar. Group 2 students stated, "The corn syrup did not go inside of the egg because the egg was floating on top of the corn syrup otherwise it would have sunk because the syrup is heavy." We consider that the students were evaluating the idea of the corn syrup pushing via a thought experiment. The students at Group 2 evaluated Group 5's theory by imagining that the egg would sink if the heavy corn syrup had diffused into it. Since the egg was still floating on top of the corn syrup, students at Table 10.2 concluded that the syrup had not moved into the egg to push the vinegar solution out. In Fig. 10.7, we refer to this as an Element Disconfirmation. The teacher, however, did not simply accept Group 5's or Group 2's statements, but asked if there was other evidence to evaluate the idea that the corn syrup entered the egg. Another group responded that the egg was deflated, as further evidence against the syrup going in.

The teacher then turned to the second idea or model element and asked the students whether the vinegar had left the egg. Students indicated that the syrup smelled like vinegar and that the top of the liquid was lighter. The teacher then conducted a demonstration activity that consisted in extracting with a straw a small amount of liquid from the top of the jar and another small amount of liquid from the bottom of the jar. She put the two liquid samples in a Petri dish for comparison. The activity enabled the students to realize that the sample taken from the top of the jar had a lower viscosity, indicating the presence of the vinegar solution mixed with syrup, and that the sample taken from the bottom of the jar had a higher viscosity, indicating the presence of pure corn syrup. We refer to these three episodes as working within an Element Confirmation Mode – to describe the type of interaction taking place and its role in model construction (Fig. 10.7).

The teacher then supported the students in explaining *why* the syrup did not go into the egg by adding two elements: (1) corn syrup was too thick to go through the membrane; (2) syrup molecules were too big. This is interpreted as a Model Modification Mode. These instances form an explanation at a deeper level for why the corn syrup did not go into the egg. For this reason, they are shown as a separate horizontal track in Fig. 10.8. In essence, they form a micro model that explains one aspect of a larger model. The teacher then conducted a fourth instance in Confirmation Mode (Fig. 10.8) by opening the deflated egg and confirming that the

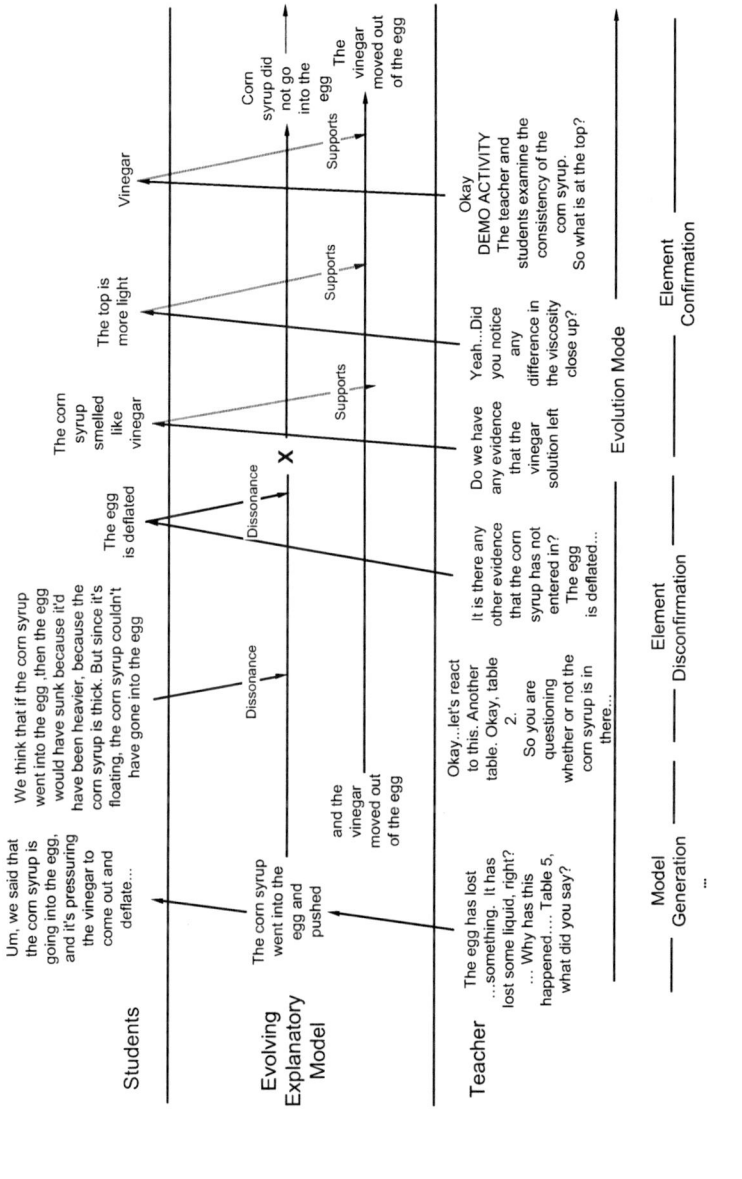

Fig. 10.7 Learning pathway showing the disconfirmation of the students' initial ideas during the construction of the concept of diffusion

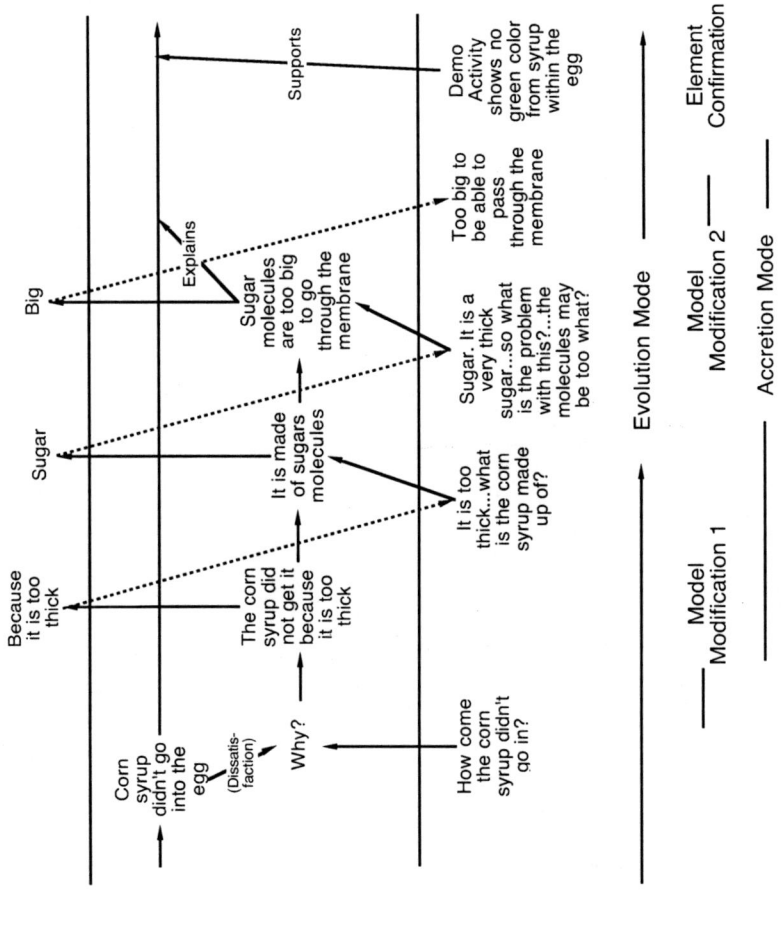

Fig. 10.8 Learning pathway showing another aspect of the construction of the concept of diffusion

syrup did not cross the membrane by finding no green coloring inside. All but the last exchange in Fig. 10.8 also fit the teacher-student interaction pattern called the Accretion Mode as defined earlier.

In summary, three teacher-student interaction patterns or co-construction modes (Disconfirmation Mode, Confirmation Mode, and Accretion Mode) were used to explain how the teacher supported the students in evaluating and modifying the first idea generated by the students: "the corn syrup went into the egg and pushed the vinegar solution to come out and deflate" (Figs. 10.7 and 10.8). In other words, the process of constructing this idea was not a single act of teaching but the consequence of skillfully combining these co-construction modes or cognitive sub-processes, which were in turn supported by specific teaching tactics (thought experiment, discrepant question, or demonstration activity).

The teacher and the students then worked to construct a different mental model to explain how the vinegar solution moved out of the egg (diffusion) (Fig. 10.9). To construct the model of diffusion, the teacher asked the groups to explain *why* the vinegar solution moved out of the shell-less egg (Fig. 10.9). The students, however, were not able to generate any fruitful ideas. The students seemed not to have a mental model of the diffusion process. The teacher conducted an activity that we call the "tea bag demo-analogy" in which she used the movement of the tea particles into hot water as a visual analogy that initiated the construction of the mental model of the diffusion process. The students observed, drew, and described how the tea particles moved from the tea bag (area of high concentration) into the hot water (area of low concentration).

The teacher then refined the initial model about diffusion by using an analogy that we call the "perfume analogy" (Fig. 10.9). The teacher asked the students to imagine an open perfume bottle and the perfume molecules spreading through the air around the room from an area of high concentration to an area of low concentration. They said that persons closer to an open bottle of perfume would be able to smell the scent before persons further away from the bottle. We hypothesize that the "perfume analogy" contributed the addition of imagery for constructing the mental model of the diffusion process. Finally, the teacher reviewed the process by explaining why the vinegar solution moved outside of the egg (Fig. 10.9). She also helped them to reestablish that the corn syrup did not move inside of the egg because "it's too thick", or, as she put it, because the molecules were too big to cross the egg membrane.

In summary, the teacher supported the students in inventing and refining an explanation for the observed egg system. She succeeded in stimulating students to generate a model that was at least half-correct. She stimulated other students to find the main flaw in the first model,

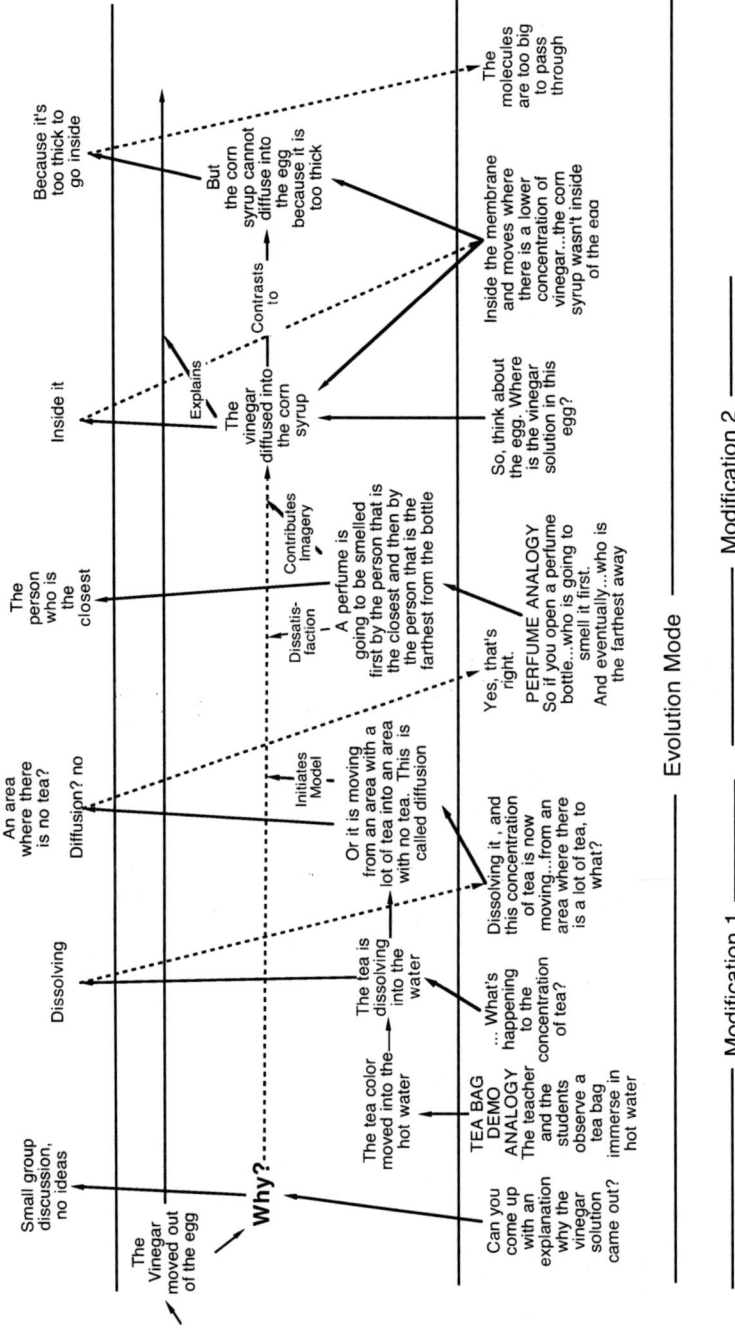

Fig. 10.9 Learning pathway showing the part of the process of constructing the concept of diffusion

encouraged them to provide further dissonance, and advocated removing that element of the model. Finally, she encouraged them to make modifications producing deeper levels of explanation in the model by asking *why* questions and using analogies.

10.4 Discussion

Throughout this chapter, our purpose has been to identify and define new concepts for describing model-based modes of teaching and learning in large classroom discussion. We have described some of the major teaching strategies used by a successful middle school science teacher while supporting her students in building mental models, including the Disconfirmation Mode, the Modification Mode, the Confirmation Mode, and the Accretion Mode. Use of these modes depends on the teacher being able to defer telling students what is wrong about their initial ideas and defer giving them the scientific model up front, so that they have a chance to participate in evaluating and modifying these models. Each process moves the student's model a small step forward through a series of successive intermediate models that we call a "learning pathway" (Scott et al., 1990; Rea-Ramirez, 1998). In this view, learning results from promoting teacher-student construction and criticism cycles that produce successive intermediate mental models until reaching the target concept.

When a string of two or more evaluation and modification cycles occurred, we identified this as a larger mode that organizes the others into what we call the Evolution Mode. This was done to express the idea that the students' model was evolving with each successive modification. Model Evolution has been referred to in other Chapters (2 and 7) as a GEM cycle of model generation, evaluation, and modification. For instance, in the first example discussed, the teacher initiated discussion by generating an initial analogical model – blood as a pasty liquid - that stimulated a student to conjecture that blood moves up because blood under it keeps pushing. The teacher then supported evaluation and revision cycles fostered by a discrepant question ("But is the blood a pasty liquid?") and the flap-in the throat analogy. These, in turn, triggered spontaneous student analogies. Thus, the teacher guided the evaluation and modification process until students reached the target concept.

The same type of Model Evolution cycle is visible in the second example. The teacher used a demonstration experiment to motivate the students to generate an explanation for why the vinegar solution moved out of the shell-less egg. The teacher then promoted evaluation and revision cycles by asking for student evaluations of the model ("let's react to this"), asking for conflicting and supporting evidence for further evaluation, requesting deeper explanations with the two different "why?" episodes shown in Figs. 10.8 and 10.9, and using analogies to refine the model (Fig. 10.9).

Gradual step-wise model evolution is perhaps the most important overall strategy in the approach presented in this book, since it can be used in so many contexts where one needs to build up a complex model. It is seen as a very different alternative to a sweeping all-at-once installation of a new model presented by the teacher. This is both because of the important contribution of the students, and because working in small "mind sized" steps means that students can follow the reasoning needed to make sense of the model. Model evolution has been documented as a major pattern in the history of scientific thinking (Darden, 1996; Nersessian 1992a, 1992b, 1995). And Clement (1989, 2008) has conducted think-aloud studies documenting the use of GEM cycles as a central process by experts when generating a new understanding about a topic. Therefore, this teaching approach models an important thinking pattern in real science.

We can break down each of the processes in the GEM cycle of model evolution as follows: Model generation at the beginning of each of the two transcript sequences was the result of a teacher-generated analogy in one case and a student-generated explanation in the other. Model evaluations were fostered through either a Confirmation Mode or Disconfirmation Mode. On other occasions, dissatisfaction was implicitly produced by the teacher's suggestion of a new element or analogy. Model modifications resulted in subtracting, replacing, or adding a new idea. These modifications occurred from either subtractions of an element from the model as the result of disconfirming evidence, replacements of an element to repair a dissonance-producing feature, or the addition of an element, usually in response to a *"why"* question.

10.4.1 Nested Structure

The categories and diagrams in this chapter were challenging to construct because the processes have a nested structure. Two kinds of nesting are shown in the diagrams: nesting of modes and nesting of explanations. Nesting of modes occurs in the way the generation, disconfirmation, confirmation, and modification modes take place within the model evolution mode, with the latter mode stretching throughout the whole sequence (as opposed to model competition which was the dominant mode in Chapter 7). Nesting of explanations occurs in the way the shriveling of the egg is explained by the vinegar leaving, and that in turn is explained at a deeper level by diffusion. Both kinds of nesting appear in the diagrams using two different notations. The diagrams in this chapter were originally designed to highlight model evolution over time and the contributions of both the teacher and the students in a process of co-construction. Therefore, we are encouraged that they can also be used to show nested processes and that discussions at this level of complexity can be analyzed.

10.5 Conclusion

The goal of this chapter was to account for two relatively complex instructional sequences involving multiple, nested modes of teaching to promote conceptual change. We have attempted to uncover certain implicit strategies that a teacher used to draw out student arguments and models. The development of the models was the result of the joint effort between the teacher and the students. It was not the product of a sudden insight but a slower step-by-step, model evolution process. Even though the teacher's goals were primarily content-driven in this case, these lessons also speak to many process goals. In fact, the students appeared engaged in a number of processes that correspond to science process goals: generating a model, evaluating a model by disconfirming or modifying model elements, modifying a model, and progressing to a deeper level of explanation. The diagrammatic notation developed is capable of showing each of these processes and the contributions made to them from both the teacher and students. We believe these "co-construction modes," shown at the bottom of Figs. 10.7, 10.8, and 10.9, are general modes of instruction that will be recognizable in analyzing other transcripts, and that teachers should be aware of them as alternative ways to interact with students in scientific discussions. The modes are conceptual tools for describing different approaches that teachers use to foster conceptual change. They have been examined from the perspective of a theoretical framework of model construction theory, with model evolution via co-construction as the central overall strategy. We believe that this framework provides a set of lenses that complements other cognitive and sociological frameworks for analyzing classroom discussions.

References

Clement, J. (1989). Learning via model construction and criticism. In G. Glover, R. Ronning, & C Reynolds (Eds.), *Handbook of creativity: Assessment, theory and research.* (pp 341–381). New York: Plenum.

Clement, J. (2000). Analysis of clinical interviews: Foundations and model viability. In A. E. Kelly, & R. Lesh (Eds.), *Handbook of research methods in mathematics and science education.* (pp. 547–589) Mahwah, NJ: Lawrence Erlbaum.

Clement, J., & Steinberg, M. (2002). Step-wise evolution of mental models of electric circuits: a "learning aloud" case study. *Journal of the Learning Sciences 11*, 389–452.

Clement, J., (2008). *Creative model construction in scientists and students: Imagery, analogy, and mental simulation.* Dordrecht: Springer.

Darden, L. (Ed.) (1996), *PSA 1996, Proceedings of the 1996 Biennial Meeting of the Philosophy of Science Association*, Part I: Contributed Papers, *Philosophy of Science*, Supplement to Vol. 63.

Duschl, R. A., & Ellenbogen K. (1999). Middle school science students' dialogic argumentation. In M. Komorek, H. Behrendt, H. Dahncke, R. Duit, W. Graeber, & A. Kross (Eds.), *Research in science education – past, present, and future* (Vol 2, pp. 420–422) Kiel, IPN Kiel.

Duschl, R. A., & Osborne, J. (2002). Supporting and promoting argumentation discourse in science education. *Studies in Science Education, 38*, 39–72.

Glaser, B. G., & Strauss, A. I. (1967). *The discovery of grounded theory.* New York: Aldine Publishing Company.

Hatano, G., & Inagaki, K. (1986). Two courses of expertise. In H. Stevenson, H. Azuma, & K. Hakuta (Eds.), *Child development and education in Japan* (pp. 262–272). New York: W. H. Freeman and Company.

Kelly, G. J., Druker, S., & Chen, C. (1998). Students' reasoning about electricity: combining performance assessments with argumentation analysis. *International Journal of Science Education 20*, 849–871.

Nersessian, N. J. (1992a). How do scientists think? Capturing the dynamics of conceptual change in science. In R. N. Giere (Ed.), *Cognitive models of science.* (pp. 3–44), Minneapolis: University of Minnesota Press.

Nersessian, N. J. (1992b). In the theoretician's laboratory: thought experimenting as mental modeling. *Philosophy of Science Association 2*, 291–301.

Nersessian, N. J. (1995). Should physicists preach what they practice? Constructive modeling in doing and learning physics. *Science & Education 4*, 203–226.

Nunez-Oviedo, M. C. (2003). Teacher-student co-construction processes in biology: strategies for developing mental models in large group discussions. Doctoral Dissertation University of Massachusetts at Amherst MA.

Nunez-Oviedo, M. C., Rea-Ramirez, M. A., Clement, J., & Else, M. J. (2002). Teacher-student co-construction in middle school life science. *Proceedings of the AETS Conference.*

Osborne, J., Erduran, S., Simon, S., & Monk, M. (2001). Enhancing the quality of argument in school science. *School Science Review 82*: 63–70.

Rea-Ramirez, M. A. (1998). *Models of conceptual understanding in human respiration and strategies for instruction.* DAI - 9909208, Amherst: *University of Massachusetts*

Rea-Ramirez, M. A. (2005). *Using imagery to facilitate reasoning with a model.* Paper presented at AERA. Montreal, Canada.

Rea-Ramirez, M. A., Nunez-Oviedo, M. C., Clement, J., & Else, M. J. (2004). *Energy in the human body curriculum. National Science Foundation.* University of Massachusetts at Amherst.

Rea-Ramirez, M. and Clement, J. (1998). In search of dissonance: The evolution of dissonance in conceptual change theory, *Proceedings of National Association for Research in Science Teaching.*

Resnick, L., Salmon, M., Seitz, C. M., Wathen S. H., & Holowchak, M. (1993). Reasoning in conversation. *Cognition & Instruction 11*, 347–364.

Scott, P. H. (1992). Pathways in learning science: A case study of the development of one student's ideas relating to the structure of matter. In R. Duit, F. Goldberg & H Niedderer (Eds.), *Research in physics learning: Theoretical issues and empirical studies. Proceedings of an International Workshop held at the University of Bremen* (pp. 203–224). Kiel, Germany: IPN/Institute for Science Education.

Toulmin, S. (1972). Human understanding: an inquiry into the aims of science. Princeton, NJ: Princeton University Press.

Chapter 11
Role of Discrepant Questioning Leading to Model Element Modification

Mary Anne Rea-Ramirez

Western Governors University

Maria Cecilia Núñez-Oviedo

Universidad de Conception

11.1 Introduction

That questioning is an important aspect of good teaching has been well established. However, the impact that specific strategies have on cognitive processes that stimulate mental model construction is less well researched. The role that one particular strategy, discrepant questioning, has on stimulating model construction is investigated in teaching systems of the human body to seventh grade students. Transcript data and drawings from case studies suggest that discrepant questioning can play an important role in stimulating dissatisfaction that leads to model element modification. Five stages in this learning process are outlined along with implications for teacher education.

11.2 Role of Discrepant Questioning Leading to Model Element Modification

Research on the importance of questioning as a teaching strategy is well documented. It is suggested that up to fifty percent of a teacher's time might be spent on questioning (Cotton, 1988) and that teachers ask between 300–400 questions a day (Leven & Long, 1981). Of the hundreds of questions teachers ask in one class period, it is estimated that approximately 60% of these questions require factual recall, about 20% require the use of higher cognitive processes, and the remaining 20% are procedural. (Gall, 1970, 1984; Hassard, 2004). According to Cotton (1988) the extensive use of questioning as a strategy in classroom teaching and the potential for effecting student learning, has stimulated researchers to look particularly at how questioning methods might affect student achievement. In addition, many studies have been conducted on the type of questions asked, wait time after asking the question, placement and timing of the question (Atwood and Wilen, 1991; Rowe, 1972; Stahl, 1994; Tobin, 1987), the reason for using various questions (Brualdi, 1998; Gooding, Swift, Schell, P. R. Swift & McCroskery, 1990), and use of questions to motivate students (Chuska, K., 1995). Some researchers have especially focused on the questions asked by the teacher during the instruction (Bean, 1985; Ellis, 1993; van Zee & Minstrell, 1997; van Zee, Corey, Minstrell, Simpson, & Stimpson, 1992; van Zee, Iwasyk, Kurose, Simpson, & Wild, 2001). Where these studies have used sociolinguistics and sociocultural perspectives to conduct the analyses, others have examined the questioning strategies through the lens of model construction and criticism theory (Clement, 1989, 1993; Craik, 1952; Johnson-Laird, 1983).

We have investigated the role that one particular strategy, discrepant questioning, has on stimulating optimal dissatisfaction with subsequent model element modification. We refer to model element to differentiate it from the entire mental model, i.e. structure of the alveoli as opposed to the whole lung structure. The research described in this paper is part of a larger project with an initial goal of understanding how students construct mental models in biology and then to develop a research-based curriculum to instruct the students about human respiration by using model construction and criticism theory (Clement, 1989, Rea-Ramirez, 1998). The Energy in the Human Body Curriculum (Rea-Ramirez, Nunez-Oveido, Clement, & Else, 2004) is organized around five individual target models that are related to the digestive system, the microscopic structure of the cells, internal structure of the cell, the circulatory, and the pulmonary systems. Rea-Ramirez (1998) conducted the initial steps to develop the curriculum including the detection of student alternative conceptions with regard to human respiration by using a sample with n = 358. The process continued with individual tutoring interviews of 12 students to determine how students were constructing understanding and what the most effective strategies were to promote model construction. These strategies were then tested in multiple teaching sessions with a small group of four students to determine their effectiveness when teaching more than one student simultaneously and the effect of the interaction between group members on mental model construction. Finally, the curriculum underwent extensive whole classroom trials in three schools over the four years and has now been developed into an interactive online curriculum. While significant gains in understanding overall were measured in these last trials (Rea-Ramirez, 2003), in this paper we deal only with the role of discrepant questioning. Examples are drawn from the tutoring studies with 12 individual students and the subsequent study of a small group of 4 students. Our purpose in presenting these case study examples is to show how it is possible to document and study the use of discrepant questions in instruction and to analyze the learning processes they can produce.

We refer to conceptual change as any significant conceptual growth, not necessarily those requiring unlearning prior to the development of a new model, nor requiring a developed prior conception. We refer to preconceptions as those ideas that the student brings to the learning situation. They may be either *naïve*, that is, those conceptions students hold that, while not incompatible with currently accepted theory, are simplistic and not scientifically complete from an expert's view, or *alternative* (sometimes called misconceptions in the literature), those conceptions that are incompatible with currently accepted scientific conceptions (Rea-Ramirez, 1998). This is an important distinction in biology where the abstract and often hidden nature of the topics often causes students to harbor either naive or alternative conceptions of complex topics. For this paper we refer to all of these conceptions as preconceptions.

11.3 Use of Questioning in Mental Model Construction

Research on the Energy in the Human Body curriculum at the University of Massachusetts (Nunez-Oviedo, 2001; Rea-Ramirez, 1998) found that different styles of questions, initially referred to simply as probing questions, were used in the to facilitate mental model construction and to more deeply probe student understanding. These questions occurred throughout the lessons, in many cases appearing spontaneously in response to a need to facilitate student mental model construction. Further analysis, however, suggests that these questions played a much larger role in the teacher-student co-construction of mental models. This is consistent with Glynn and Duit's (1995) contention that if questions are asked during and after a lesson they could encourage students to evaluate, revise, generalize, and apply their knowledge.

Questions were then divided into two major types, "supporting questions" and "discrepant questions." Supporting questions were employed throughout the curriculum to activate students' existing knowledge, relate this knowledge to experiences, and intrinsically motivate students (Rea-Ramirez, 1998). In most instances supporting questions were used in the generative and modification phases of conceptual model development. They were not intended to produce dissonance but rather to assist the student in generating ideas before and after the evaluation phase.

11.4 Discrepant Questioning

When students generate a model element that is not consistent with the target model or the ongoing discussion, two options are to: (1) tell the students the scientific model, or (2) use a strategy to cause dissatisfaction within the student concerning their model. A discrepant question is one designed to produce dissatisfaction with a student's model or conception (Gibson & Rea-Ramirez, 2002), Rea-Ramirez and Nunez-Oviedo, 2002, Clement, 2002. This type of questioning is very different from those that cause accretion (the result of a series of leading questions that cause students to generate one small element of the model at a time through easy inferences or prior knowledge). We believe it is important to make a clear distinction and to suggest the use of more specific terminology (Nunez-Oviedo & Rea-Ramirez, 2007). In this paper we will deal only with discrepant questioning.

While Cotton (1988) defines teacher questioning as "instructional cues or stimuli that convey to students the content elements to be learned and directions for what they are to do and how they are to do it," we suggest that discrepant questions go beyond simply providing students with cues to convey content elements. Cotton's definition implies that the questioning is suggesting to the student what to pay attention to, what is important in the teacher's view, and what directions to follow in learning the information. On the other hand, a discrepant question may actually stimulate students to engage in a cognitive process that is much more student driven and interactive, in which they are actually evaluating and modifying or constructing the information in the form of a workable, runnable model.

It is also important to distinguish discrepant questioning from discrepant events. Thompson (1989) states that a discrepant event occurs "when something a child expects to happen does not occur. The result is the opposite of what was expected, and it contradicts the belief of the individual," thereby referring to an observed event. In contrast, discrepant questioning does not involve actual events or experiments but puts forth an idea or concept (or possibly a thought experiment) that is in contrast to a belief held by the student, generally in the form of an open-ended question. But, as, Thompson suggests, like discrepant events, a discrepant question "throws the child off balance intellectually and which most likely will motivate this person to further investigate the science concept. However, since much of the reaction to the discrepant question and the initial response is internal to the student, we found it necessary to use transcript microanalysis to develop a greater understanding of the effect and use of such questioning.

11.5 Modeling the Effects of Discrepant Questioning

In this section we present our theoretical framework for how a discrepant question can cause conceptual change through modification. This framework is now partly a product of analyzing the transcripts to be presented but we present it here as an advanced organizer. Figure 11.1 graphically depicts the sequence of model element change that occurs in response to a discrepant question. We hypothesize that students attempt to reason about the discrepant question in conjunction with running a mental model that they already have. When the result is not compatible with the context of the question, students experience dissatisfaction. This can motivate them to form a new idea or modification of the initial mental model element. Fostering building of the student's model is an important element of the teaching process, stressing that it is more than just causing the student to be dissatisfied. When model element change is incorporated into, or gives rise to, a model that is more like the target, as intended by the instructor, we call this normative change. Change is often not evident until the model (not just the model element) has undergone multiple intermediate model changes.

11.5.1 Examples from Transcripts

Not all components are readily evident in every instance since some of the components such as student dissatisfaction are hard to substantiate, being based often on the observation of facial expressions, hand movements, or simple utterances. Figure 11.1 graphically depicts the components of model element change that can occur in response to a discrepant question. Therefore, we base the presence of a model element change on evidence for the documented three components contained in the dark boxes in Fig. 11.1, with support from suggested evidence of the other two components.

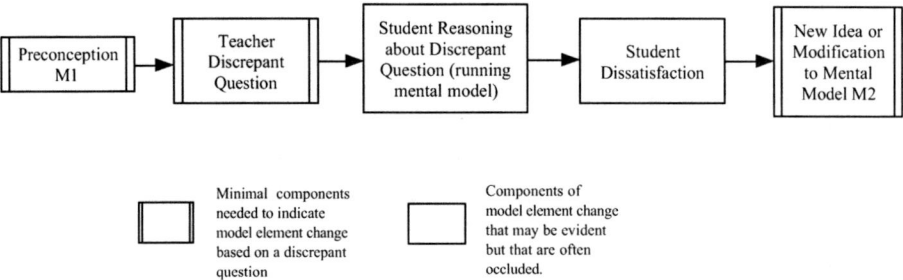

Fig. 11.1 Graphic representation of the cycle of model element change in response to a discrepant question

The data provided a rich supply of discrepant questions such as:

- (Students raise question as to whether we have one lung or two.) Does anyone remember reading about someone having an operation to remove one lung?
- (Students make drawing showing that blood is only pumped out from the heart to the body one way and goes into the tissues) Where do we get more blood to keep pumping out? Why don't you need a transfusion every now and then?

We will discuss several examples in more depth to illustrate how the teacher supported the students through the dissatisfaction and construction process to evaluate their ideas by using discrepant questions. In addition, we will discuss how various discrepant questions vary in the amount of 'leading' that they do for subsequent student reasoning.

11.5.1.1 Example One – Student I

This example was taken from the individual tutoring interviews while discussing the pulmonary system. The student held the alternative conception that the lungs were simply two small tubes connecting the throat to the heart. During the first phase, the teacher initially asked students to draw their ideas of how the body received and used air. Student I's initial drawing showed tubes sending inhaled air directly to the heart but drew lungs as separate organs unconnected to the inhaled air. In response, the tutor used a discrepant question to promote dissatisfaction and to stimulate criticism and revision cycles in the student. She began by asking the student to take a deep breath that initiated the following discussion.

> T: "What happened?"
> S: "Lungs get bigger."
> T: (Pointing to the drawing of lungs and then the tubes connecting to the arteries), "Why do they do that if the air is going directly into the heart?" (Discrepant Question)

Recall that a discrepant question is a question that is used to produce dissatisfaction with an idea generated, thereby leading to disconfirmation or modification of a model element. In the transcript segment above, Student I has initially drawn tubes directing air from the outside through the throat into the heart. Around these three tubes Student I drew the outline of round shapes she indicates are lungs. The instructor, recognizing that her drawing shows no connection between what she draws as lungs and the air (since the air remains in the three tubes that go directly to the heart), uses a discrepant question to help focus her attention on this model element in an attempt to cause the student dissatisfaction with her model and possibly a desire to modify it. The instructor uses a simple activity of having the student take a deep breath and describe the sensation. Student I's dilemma comes when she tries to reconcile her sensation of the lungs getting bigger as she inhales with her drawing that shows no connection between the outside air and the lungs. The student, then, [giving a confused look as evident on the videotape], tried to construct a mechanism. Her first indication of a model element change was that she suggested "Maybe there's lots of tubes that *expand* (uses hand gesture to show getting bigger) not just three" (she draws in more tubes). This would account for the "lungs getting bigger". While this model is still a distance from the target, it indicates a positive stepwise change in the model.

If we analyze this example using Fig. 11.1, we see that it meets the requirements for model element change, with at least three of the components present. The student has drawn and verbalized a preconception (M1) as tubes (3) directing air from the outside to the heart. The instructor then initiated a discrepant question. In response to this, the student shows in her discussion and drawing a modification of the model element. In addition, there is some suggestion that the student is attempting to reason using her initial model as stimulated by the discrepant question because she builds on her multiple tubes idea. This is evident from her comment above,

"Maybe there's lots of tubes that *expand* (uses hand gesture to show getting bigger) not just three" (she draws in more tubes) Although it is more difficult to substantiate, we also observed that she exhibits dissatisfaction by pausing with a confused look, just before modifying of her model. Thus we have some evidence for five of the boxes in the sequence in Fig. 11.1 for this case. This student's change continued as she first integrated her models of tubes and hollow lungs into one model with lungs containing many small tubes. Only later, after several other criticism and revision cycles did she construct alveoli to go with the tubes and even later a capillary system.

11.5.1.2 Example Two – Student VI

While discussing the same topic, Student VI struggled when asked to draw her concept of how oxygen got into the blood stream from the air. In this instance the student had the alternative conception that there is a direct connection from the mouth through the throat to the heart. She stated, "You breathe in oxygen and it goes to the heart... Maybe it goes to veins and veins take it to the heart. From the heart it is pumped out in the blood through the blood stream." Her oral description of the drawn model did not actually include the lungs as a necessary part, similar to that of Student I, even though structures she labels as lungs are drawn. The tutor attempted to introduce dissatisfaction as part of the Evaluation phase by asking a discrepant question, "I wonder why we need lungs then?" Since Student VI made no apparent connection between the lungs she drew and the air that she showed moving from outside, through tubes (suggested to be veins), and directly to the heart bypassing the lungs, the instructor attempted to use the discrepant question to direct the students' attention to the lack of connection between the lungs and the "veins".

Again following Fig. 11.1, possible reasoning about the discrepant question, running the model, and dissatisfaction were suggested by the student's facial expression and stammering of "Ummmm——use them to (pause) that's what helps you to inhale." The tutor encourages her to explain further by asking, "How does that work?" Student responds, "when inhale, sucks in air and oxygen......ummm." While she was unable to suggest another model at this time, the above expression and stammering suggested that she was beginning to be dissatisfied with the old model. However, unlike Student I, Student IV had difficulty modifying her idea immediately. When this occurs, the role of the teacher moves from using discrepant questions that are very open ended to questions that may be more directed or leading. In this instance, the tutor used the following discussion that included another discrepant question to help her in the modification process.

    T: How does that suck in air? Maybe if I had a clear picture of what the lung looked like.
    S: Air is in the throat. Veins go through the lungs. (Student drew a picture showing an irregular oval shape with three large tubes running diagonally from the top where it was open to the air to the lower right side where the tubes went through the wall of the oval and then stopped)
    T: Why would they (pointing to the tubes she drew) need to go through the lungs? (Discrepant Question)
    S: Goes *through* the lungs (indicating passes through). (pauses, then states) Goes into the lungs then into veins then to the heart.
    T: Why might this way be better than the first way you showed me?
    S: Then [in the first model] not just oxygen would go to the heart because you breathe in other things too. (It appears that the student is suggesting that since we need only oxygen to go through the heart to the cells, that you need some other structure to 'filter' out the other components of the air, sending only oxygen to the heart.)

    The instructor recognizes a discrepancy in the preconceived model (M1) that the student has drawn, *i.e.* that while the lungs are present on the drawing around the tubes, there is no connection between the air outside and the lungs. The air is transmitted through the tubes directly to the heart. In an attempt to encourage the student to recognize that this is model does not account for the presence of lungs, possibly having no need for them if there air is going to move directly from the outside through tubes to the heart, the instructor asks a discrepant question. Rather than simply attempting to cause dissatisfaction, the tutor, using the discrepant question, draws the student's attention to the specific problem in the model. The student responds to the discrepant question by modifying her model to suggest that the air actually goes into the lungs first and then into the tubes she calls veins to the heart. At this point the instructor asks the student to draw this new model. The student has modified her model now to include many tubes and states, " I'm not sure what it looks like. Air stays in the tubes. The lungs are just a bunch of tubes from the throat." This revised model has many tubes that fill the lungs, which is partially correct. This we refer to as model element modification. It then took multiple additional cycles involving model element revisions to arrive at a model that was close to the target. When she was asked to recreate the entire process she stated, "Oxygen comes into the throat, tubes branch and end in sacs. The capillaries pick up oxygen and take it to veins and to the heart then into the body. Carbon dioxide comes out in the reverse direction." Her initial misconception did not reappear.

11.5.1.3 Example Three – Small Group

Sometimes the model element that needs to be modified is very small and quickly changed using a discrepant questioning. This, however, does not suggest that it is less important. In constructing the lungs, students in the small group began with very different models, all of which were inconsistent with the target. Figure 11.2 shows the initial individual student lung models. The instructor, then, asked the students to draw a consensus model (Fig. 11.3). Three of the four drew either single or double, balloon like lungs. Double lungs were connected in the middle. Each showed air moving within the open center. The fourth student had a similar open area for air in the center of his drawing but drew layers of cells around it that he noted are protective.

One student was selected from the group to be the one to draw what other students suggested on a large whiteboard. The drawer was encouraged to ask questions to be sure he was including all students' ideas and to get the other students to pool their ideas into one model. In the whiteboard model, the students contributed characteristics from their initial drawings. The consensus model combined elements from each of the initial preconceptions, mostly an open area for air in the middle of the structure but added a single layer of cells around the outside and some structures they called veins coming from the cells into the open center. In the consensus model they all agreed to leave an opening at the bottom of the structure that had been included in Barbara and Martha's models and in Gary's model on the side (All names have been changed to protect the students). Louis used a tube instead of an opening in his model to provide an opening between the inner open area and the outside of the lung.

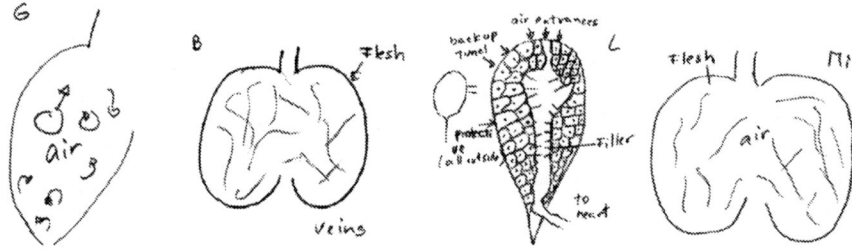

Fig. 11.2 Initial pre-conceptual models of lungs

Fig. 11.3 Consensus model of lungs showing an opening at the bottom where air can escape

In order to promote model construction, the teacher accepted the students' ideas and asked students to help her in understanding the structures and connections included in the consensus model. The teacher, however, found several problems in this model and used discrepant questions and other tactics, such as discrepant events, and analogies, to stimulate dissatisfaction leading to modification. Through each of these criticism and revision cycles, the teacher supported the students to solve one problem at a time until examining all of the aspects of the students' ideas that were not compatible with target model of the lungs.

The first problem encountered in the consensus model was the presence of a hole at the bottom of the lungs (Fig. 11.3). To support the students in modifying this model element, the teacher asked the students a discrepant question.

Teacher: I have a little question here because you have this sort of space here (points to hole in bottom of lung) that all the air can sink out of, what do you want to do about that? (Discrepant Question)

Martha: Close it up
Gary: Close it up
Louis: You want to close it up right there. (student at whiteboard points to hole in the bottom of the lungs).
Teacher: Why?
Louis: Why?
Martha: Because you need to hold the air in your lungs and comes back out the same place that came in.
Gary: Because…
Martha: Where else is it going to go?
Louis: I do not know.
Martha: [Breathing deeply and feels chest movement]

Having accessed the students' preconceptions (M1) as drawn in Fig. 11.3, the instructor draws the students' attention to this problem by using a discrepant question that is both directed and leading. In response, students appeared to recognize, almost immediately, that air doesn't just sink out of the lungs, even though they had not yet devised an effective way for air to move from the lungs out to the body. There is evidence that suggests reasoning about the discrepant question when some students question why they need to close up the hole and Martha explains that you need to hold air in your lungs. This led to a model element modification that closed the hole. Note that the modified model is still basically a balloon lung model with a hollow sac holding air but without the hole at the bottom. Later, after other criticism and revision cycles leading to more model element modification, the four students arrived at a new model (Fig. 11.4) that includes lungs filled with air cavities and blood vessels in close proximity. Evidence of this is supplied in their final drawings.

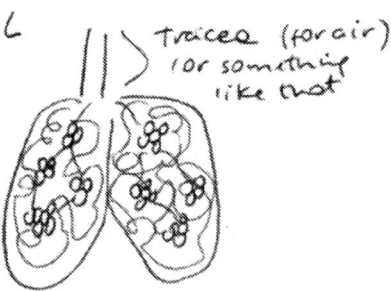

Fig. 11.4 Intermediate group model of the lungs

The discrepant question was intended to cause dissatisfaction and modification of one model element only (hole at bottom of the lung drawing). As a result, because students had now closed the lungs except from the upper end coming from the mouth, it set the stage for the teacher to use this model later to suggest that there must be another way for air (oxygen) to move from the lungs out into circulation to the cells. Therefore, although the intermediate model of a closed hollow lung that the students drew was not scientifically accurate yet, it served to modify one element in a long chain of model elements that were generated, evaluated and modified along the way to a complete mental model.

11.5.1.4 Example Four – Small Group

In the small group, while the students were discussing the location of blood vessels for the small intestine, a student suggested that blood vessels might be inside the small intestines. Therefore, the alternative conception consisted of blood vessels in the lumen of the intestines rather than being situated within the villi. The teacher presented the student's idea to the group for discussion and asked a discrepant question with regard to the presence of blood within the intestine.

> Teacher: ... (Drew picture of a segment of small intestines that resembled that of the student, with the blood vessel coming directly into the inner area of the intestines and abruptly ending there.) Okay, how do you think the blood vessel has to be in relation to the intestine in order for this stuff to move out of here, enter the blood vessels, and to go to your big toe to that cell.
> Barbara: How?
> Teacher: What does it look like, where is it going to be? Here is your intestine where should I put the blood vessel.
> Barbara: Like connecting.
> Teacher: Right here? (points to a point on the outside of the intestine)
> Gary: While traveling, isn't blood in the intestine too or something?
> Teacher: Um, I wonder when he says if there is blood in the intestine what do you think about that? (Discrepant Question)
> Louis: Oh
> Gary: I am not I am not saying that there is but [inaudible]
> Martha: Is there? I do not know.
> Teacher: Okay? What do you think?
> Louis: That would mean that blood would be coming out. I mean that every time that you go to the bathroom, you will be bleeding.

After the teacher asked the students the discrepant question, she encouraged the students to present their ideas with her question "what do you think". One of the students realized that the idea suggested by the other student simply could not work. We refer to this as producing *disconfirmation* because the model did not reemerge in the students discussions from that point forward. Students then focused on other models for circulation to the intestines. Therefore, sometimes a discrepant question led to modification of the model element causing students to restructure or refine the element, and sometimes it led to disconfirmation in which the group discarded an element that they did not find fruitful.

11.6 Discussion

Discrepant questioning as a strategy for addressing students' misconceptions is more specific than those addressed in previous literature. A broader approach, cognitive conflict, has been discussed in the literature as a basis for strategies for promoting conceptual change. This includes Cosgrove and Osborne's (1985) Generative Learning Model, Champagne, Gunstone, and Klopfer's, (1985) Ideational Confrontation Model, Chinn and Brewer's (1993) use of anomalous data. In addition, Nussbaum and Novick (1982) stress that it is important to create conflict in order to guide conceptual restructuring. While all of these authors suggest the need for causing dissatisfaction with a prior model in order to initiate conceptual change, they do not discuss the use of discrepant questioning. Sometimes, anomalous data or contrasting scientific models are presented in the hope that they will cause conflict. While this has been shown to be effective in some instances, the approaches often necessitate setting up an elaborate demonstration. In the present examples a simple short discrepant question was introduced that caused rapid dissatisfaction and modification of a model element. This suggests that it is not always necessary to use demonstrations with special equipment, but that instances of discrepant questions can be injected into the current curriculum with positive benefits. These can be generated on the spot by a skilled teacher.

Processes involved in teaching via discrepant questioning include drawing out prior conceptions, selecting or generating a question, and also detecting and supporting student reasoning about the discrepant question and the dissatisfaction that ensued, as shown in Fig. 11.1. From a researcher's point of view these last two elements are not as easy to document, as they may take place internally with no visible output. However, suggested facial expressions, attempts to draw including erasures and changes in the drawing, and verbal utterances in the examples suggested that both of these components are occurring in many of the examples. These same elements can be used by the teacher to assess whether the student is experiencing dissatisfaction and criticizing their model.

11.7 Conclusion

Through the qualitative case studies in this chapter we have endeavored to show how it is possible to document and study the use of discrepant questions in instruction and to analyze the learning processes they can produce. The cases suggest that discrepant questioning can play an important role in stimulating model element modification. They also illustrate how evidence can be gathered for stages in this learning process pictured in Fig. 11.1. The teacher first gathers information on the student's initial model or preconception before choosing a question that can cause the student to reason with that model in a way that generates dissatisfaction with it. This provides motivation for modifying the model or forming a new model. Although evidence for some of these stages can be occluded in any particular protocol, evidence for three of the stages – a preconception or initial model, a discrepant question, and a significant change in the model – seems sufficient to infer that a discrepant question has had an effect.

In the examples cited, discrepant questioning was part of a larger model evolution strategy of starting from the students model or conception of a biological system and taking it through successive cycles of evaluation and revision. The intention is that the student participates actively in the construction of a new model and thereby better understands its functioning and the reasons for why the system is structured as it is. This is done in small successive steps of criticism and revision in order to enable students to reason in manageable steps rather than asking them to make a huge leap. Sometimes successive discrepant questioning cycles were necessary to cause modification in a persistent preconception while at other times only one discrepant question was necessary for disconfirmation or modification of a model element.

11.8 Implication and Suggestions for Teachers

Discrepant questioning is a teaching strategy designed to help students focus on their own mental model of a concept, critically question their conceptions, including naïve ideas or gaps in understanding, "unlearn" misconceptions, make rational decisions about what they believe, and/or restructure initial conceptions leading to deep understanding of a single conceptual element or model. It requires teachers to question students in a way that stimulates them to examine their ideas or models, without giving information prematurely to the student or passing judgment on the student's model. It is not fact finding or accretion.

To employ discrepant questioning effectively, there are several key elements that teachers need to understand. First, many discrepant questions do not generally appear explicitly in the curriculum teachers use but rather are necessarily designed by the teacher in response to the students' models. This suggests that the teacher must be able to draw out the students' own models and understand or envision what intermediate model elements might need to be encouraged in order to arrive at a more complex mental model. This necessitates having an adequate knowledge of the content being studied

Second, the teacher should not provide the target or scientific model until after students have accessed their prior conception and struggled to construct new or modified model elements. For example, teachers may attempt to give students a scientific model of the lungs with bronchi, alveoli, and capillary circulation, by showing the scientific model at the beginning of instruction either through classroom lecture, pictures, or an actual model. However Rea-Ramirez (1998) found that students who had been presented with these models in prior school situations did not have a mental model that was deep or explanatory and that they were not able to express such a model any better than students who had no prior experience. Providing students with the scientific model first might cause students to accept the model presented by the teacher as the "right" answer, but at the same time retain alternative models that could prove to be persistent later.

Third, the teacher should focus on each individual part of the model (model elements) rather than assuming that the whole model needs to be modified at once. This requires that the teacher examine the prior conception for individual model elements that are not compatible with the target model. Initially focusing on the trying to "fix" the whole lung model at once in example 3, rather than concentrating on one element at a time, such as on the hole in the bottom, may have caused the teacher to bypass this and other alternative conceptions that needed attention Thus, it is important for the teacher to first understand the elements of a target model, compare this to the elements presented in the students initial conception, and then undertake questioning and other strategies to help the students through successive criticism and revision cycles of each model element, gradually building the whole model.

Fourth, the teacher should be willing to allow intermediate model elements to be constructed that are not yet compatible with the target model. Since students need to build the model one element at a time, often the new element that is modified or constructed is naive or incomplete until all of the elements are constructed and combined. This was the case with the hole in the bottom of the lung. While the students closed the hole it still did not cause the model to be complete or "correct". It is often difficult for teachers to allow partial models to be considered in this manner. However, this is a necessary condition to enable students to construct the model in small successive steps.

Finally, teachers should recognize that the use of discrepant questions does not always have to be time consuming. The length of time of the interaction resulting from a discrepant question depends on the size and nature of the model element. And it should not be used on its own but rather alongside other strategies. For this reason, teachers should not consider the use of discrepant questioning as a new pedagogy that will replace other currently used strategies, but rather an additional tool to be integrated into their existing curriculum.

11.9 Suggested Further Research

Further studies of both university and public school teacher uses of discrepant question should be conducted to provide additional insights into how teachers can use this strategy and also to suggest possible questions that can be integrated into curriculum. In addition the use of discrepant questioning may be a useful strategy for instructing pre-service teachers. Davis (2001) suggests that conceptual change "is not only relevant to teaching in the content areas, but it is also applicable to the professional development of teachers and administrators.... Teachers must learn different instructional strategies, but they must also reconceptualize or change their conception about the meaning of teaching." Pilot studies are currently being conducted by the authors in both traditional and online universities to investigate the role of discrepant questioning in teacher preparation.

References

Atwood, Virginia A., & William W. Wilen. (1991). "Wait time and effective social studies instruction: What can research in science education tell us?" *Social Education, 55*, 179–81.
Bean, T. W. (1985). Classroom questioning strategies: Directions for applied research. In A. C. Graesser & J. B. Black (Eds.), *The Psychology of Questions* (pp. 335–358). Hillsdale, NJ: Lawrence Erlbaum Associates.
Brualdi, A. (1998). *Classroom questions.* Washington, DC: ERIC Clearinghouse on Assessment and Evaluation. [ED422407]

Clement, J. (1989). Learning via model construction and criticism. In C. Reynolds (Ed.), *Handbook of creativity: Assessment, theory and research* (pp. 341–381). New York: Plenum.

Clement, J. (1993). *Model construction and criticism cycles in expert reasoning.* Paper presented at the Fifteenth Annual Meeting of the Cognitive Science Society, Hillsdale, NJ.

Champagne, A. B., Gunstone, R. F., & Klopfer, L. E. (1985). Effecting changes in cognitive structures among physics students. In L. West & A. Pines (Eds.), *Cognitive structure and conceptual change* (pp. 163–188). Orlando, FL: Academic Press.

Chinn, C. A. & Brewer, W. F. (1993). The role of anomalous data in knowledge acquisition: A theoretical framework and implications fro science instruction. *Review of Educational Research, 63*(1), 1–49.

Chuska, K. (1995). *Improving classroom questions: A teacher's guide to increasing student motivation, participation, and higher-level thinking.* Bloomington, IN: Phi Delta Kappa Educational Foundation.

Cosgrove, M., & Osborne, R. (1985). Lesson frameworks for changing children's ideas. In R. Osborne & F. P. Freyberg (Eds.), *Learning in science: The implications of children's science* (pp. 101–111). Portsmouth, NH: Heinemann.

Cotton, K. (1988). *Monitoring student learning in the classroom.* Portland, OR: Northwest Regional Educational Laboratory.

Davis, J. (2001). Conceptual Change. In M. Orey (Ed.), Emerging perspectives on learning, teaching, and technology. *http://www.coe.uga.edu/epltt/conceptualchange.htm.*

Ellis, K. (1993). *Teacher questioning behavior and student learning: What research says to teachers.* Paper presented at the 1993 Convention of the Western States Communication Association, Albuquerque, New Mexico. (ED 359 572).

Gall, M. (1970). The use of questions in teaching. *Review of Educational Research, 40,* 707–721.

Gall, M. (1984). Synthesis of research on teachers' questioning. *Educational Leadership, 42,* 40–47.

Gibson, H., & Rea-Ramirez, M. A. (2002). Keeping the inquiry in curriculum designed to help students' conceptual understanding of cellular respiration. *Proceedings of the AETS Conference.*

Glynn, S., & Duit, R. (1995). Learning science meaningfully: Constructing conceptual models. In S. Glynn & R. Duit (Eds.), *Learning Science in the Schools.* Mahwah, NJ: Erlbaum Associates.

Gooding, J. N., Swift, Schell, P. R., Swift, & McCroskery, (1990). *Journal of Research in Science Teaching, 27,* 789–802.

Hassard, J. (2004). Online excerpt from: *The art of teaching science.* Oxford University Press. URL: http://www.scied.gsu.edu/Hassard/mos/Johnson-Laird, P. N. (1983). *Mental models.* Cambridge, MA: Harvard University Press.

Leven, T. & Long, R. (1981). *Effective instruction.* Washington, DC: Association for Supervision and Curriculum Development.

Nunez-Oviedo, M. C. (2001). *A teaching method derived from model construction and criticism theory.* Unpublished Comprehensive Exam Paper, University of Massachusetts, Amherst.

Nussbaum, J., & Novick, N. (1982). Alternative frameworks, conceptual conflict, and accommodation: Toward a principled teaching strategy. *Instructional Science, 11,* 183–200.

Rea-Ramirez, M. A. (1998). Model of conceptual understanding in human respiration and strategies for instruction. (Doctoral dissertation, University of Massachusetts, Amherst, 1998). *Dissertation Abstracts International, 9909208.*

Rea-Ramirez, M. A., & Nunez-Oviedo, M. C. (Jan 2002). D*iscrepant questioning as a tool to build complex mental models of respiration.* Paper presented at AETS, Charlotte, NC.

Rea-Ramirez, M. A., Nunez-Oviedo, M. C., Clement, J., & Else, M. J. (2004). *Energy in the human body curriculum.* University of Massachusetts, Amherst.

Rea-Ramirez, M. A., & Else, M.J. (Mar, 2003) Evidence of model based reasoning in human respiration. *Proceedings of NARST 2003 Conference, PA.*

Stahl, R. J. (1994). *Using "think-time" and "wait-time" skillfully in the classroom.* ERIC Clearinghouse for Social Studies/Social Science Education Bloomington IN. ED370885

Thompson, C. (1989). Discrepant events: What happens to those who watch? *School Science and Mathematics, 89*(1), 26–27.

Tobin, K. (Spring 1987):. "The Role of Wait Time in Higher Cognitive Level Learning." *Review of educational research, 57,* 69–95.

van Zee, E., & Minstrell, J. (1997). Using questioning to guide student thinking. *The Journal of the Learning Sciences, 6*(2), 227–269.

van Zee, E. H., Corey, V., Minstrell, J., Simpson, D., & Stimpson, V. (1992). *Student questioning during a cognitive approach to physics instruction.* Paper presented at the NARST Conference, Boston, MA.

van Zee, E. H., Iwasyk, M., Kurose, A., Simpson, D., & Wild, J. (2001). Student and teacher questioning during conversations about science. *Journal of Research in Science Teaching, 38*(2), 159–190.

Chapter 12
Using Analogies in Science Teaching and Curriculum Design: Some Guidelines

Mary Jane Else

University of Massachusetts

John Clement

University of Massachusetts, Amherst

Mary Anne Rea-Ramirez

Western Governor's University

12.1 Introduction: Why are Analogies Useful in Science Teaching and Learning?

Analogies are comparisons that are based on similarity. In science education, using analogies may help students to understand new material by building on familiar experiences or prior knowledge. For example, many textbooks compare the cell and its parts to a factory. In the cell/factory analogy, parts of a factory such as the foreman and the boiler are compared to parts of the cell such as the nucleus and mitochondria. This analogy helps students understand that a cell has parts that perform functions such as directing cell processes and releasing the energy in food. The parts of the factory that are similar to the cell are transferred or "mapped" to the student's developing concept of the cell.

The exploration of the role of analogies in human learning has led to a rich body of theoretical, experimental, and classroom-oriented literature. Analogies have been studied as part of the processes of natural learning in children, the reasoning processes of scientists, and in the learning of science in schools (Hatano & Inagaki, 1988; Dagher,

1994; Newby, Ertmer, & Stepich, 1995; Glynn & Takahashi, 1998). Analogies may be particularly useful in the teaching of topics that are difficult for students to construct through direct experiences, such as labs and demonstrations. The structure and function of the cell is case in point, as it is impossible to directly observe cell organelles in action.

Some recent work has focused on the role of analogies in the building and revision of mental models. When analogies are used at the beginning of teaching, they may serve to help students form an imperfect preliminary model (M1) that is later modified. An analogy may also be used at critical points later in model evolution cycles to provide missing model elements. The need to provide multiple model elements may lead to the sequential use of several analogies. For example, Johsua and Dupin (1987) used three complementary analogies – water flow, trains on a track, and refrigeration – in instruction about electrical circuits. Brown and Clement (1989), Clement (1993), and Spiro Feltovich, Coulson, & Anderson (1989) found that using multiple analogies was helpful in overcoming the shortcomings of individual analogies. In our own research a river delta, water pipes, and lobster traps were all used to convey different aspects of the structure and function of vessels in the human circulatory system (Rea-Ramirez, Nunez-Oviedo, & Else, 2002).

Analogies may have limitations as teaching tools. Each analogy has elements that the teacher intends to have students focus on and use to understand the new material. Other analogy elements are different than the material to be learned. Teaching must be accomplished in such a way that students use only the appropriate portions of the analogy. Some researchers, therefore, argue that methods of using analogies should be carefully structured. Glynn (1991) has developed a "Teaching with Analogies" (TWA) model, a six-step method of presenting analogies in classroom settings. In the TWA model, students are first introduced to the concept to be learned and guided through an explanation of the familiar portion of the analogy. In subsequent steps students draw comparisons between the familiar idea and the concept to be learned and then identify places where the analogy breaks down, places where the familiar portion of the analogy is different from the new concept. Lastly, students and teacher summarize and draw conclusions about the new concept (Glynn 1991). Bulgren et al. (2000) organized analogy presentation into a table of similarities and dissimilarities.

We suggest that there are still questions to be answered about analogy use in the classroom, particularly with young learners such as middle-school students. In this chapter, we would like to review what has been written in the literature on science teaching with analogies and what we have learned from our own research with analogies by focusing on four key questions. These are:

- How do analogies fit into a model-based approach to science teaching?
- What are some ways of maximizing the learning benefits of analogies?
- What factors characterize different types of analogies and how should these be taken into account in designing a lesson that uses an analogy?
- When are analogies the "right tool" in constructivist approaches to science teaching?

This chapter will be organized as follows. First, we will identify the terminology we will use in our discussion of analogies. Next, we will briefly describe our own research trials of a mental model-based life science curriculum. The bulk of the chapter will be our discussion of our own and others' thoughts about the questions listed above. Finally, we will conclude with a summary of what we have learned about the use of analogies in science teaching.

12.2 Terminology

We use the following terms to describe analogies and the process of using them in the classroom. We use the term *base* to describe the familiar portion of the analogy. We use the term *target* to describe the unfamiliar scientific domain, the concept that the teacher wants his or her students to learn. Analogy *elements* are pieces of the base or target that correspond to each other. *Mapping* is the process of describing relations between base and target. Lastly, we use the word *transfer* to describe the cognitive process in which understandings about the base are applied to the target.

12.3 The Energy in the Human Body Curriculum

The *Energy in the Human Body* curriculum was developed to help middle-school students understand cellular respiration and the body systems associated with it (Rea-Ramirez, 1998). Cellular respiration is the biochemical system in living things in which the chemical energy contained in the glucose molecule is released and transferred to ATP, the cell's energy "currency." The curriculum is:

- Research-based, having been developed after a set of individual and small group tutoring interviews.
- Conceptually integrated, in that cellular processes are connected to the body systems that assimilate and transport food, oxygen and the waste

products of cellular respiration. The entire curriculum forms a coherent "story" about how energy is used in the body.
- Strategic, in that multiple teaching and learning tools are used to help students build understanding, and in that these tools are employed as deemed appropriate for specific learning goals. Teaching and learning tools used in the curriculum include analogies, cooperative/small-group work, "learning by drawing," discrepant questions, recall of students' "daily life" experiences, and model generation, evaluation, and revision (GEM) cycles.
- Student-active, in that many of the ideas used in constructing knowledge come from students themselves (Rea-Ramirez, 1998).

The curriculum was assessed in middle-school classrooms in western New England for three consecutive years. Trials were conducted in five schools. Seven teachers participated at various times, with two teachers participating for the entire length of the project. Students were primarily in the seventh grade. Teachers were trained in week-long summer workshops that stressed the pedagogical principles that guided curriculum development. In addition, teachers were provided with a detailed manual that provided explanations of the pedagogical approach and described specific teaching techniques.

The curriculum, teacher training and teacher training materials were revised yearly in response to what we learned from our classroom observations and from our assessments of student learning. Observations were conducted on nearly a daily basis in two of the schools in the first two years of the trial and with decreased frequency in the last trial year. Assessments took several forms. A multiple-choice test that was administered both before and after each curriculum trial was used to determine gains in content knowledge. An open-response test, also conducted both prior to and after instruction, was used to determine students' abilities to synthesize and integrate what they had learned. Lastly, in some classrooms we added assessments administered before and after the use of specific analogies in order to gain information about the effectiveness of those analogies.

12.4 The Use of Analogies in the Energy in the Human Body Curriculum

Table 12.1 shows examples of analogies used in the curriculum. Our analogies vary both in complexity and in purpose. Complex analogies such as the school analogy have a number of elements that correspond or "map" between base and target. Simpler analogies, such as the "ear of corn" analogy, have only a few elements that map. These two analogies also vary in purpose, with the school analogy being designed to help students understand the functions of cell parts and the relations among them, and the ear of corn analogy being designed simply to generate a visual or geometric model that could help students understand how cells are arranged. In our discussion of analogies, we will suggest that characteristics of analogies such as complexity and purpose are important. We will also suggest ways to help students become aware of important analogy features.

Table 12.1 Examples of analogies used in the "Energy in the Human Body" curriculum

Analogy	Mapped elements	Complexity	Purpose/function
Ear of corn	Arrangement of kernels is like arrangement of cells in body in that both are arranged in a pattern with little space in between	Simple	Visual/geometric
School analogy	Schools have parts that have different functions, just as cells have parts that function differently, and the functions of some school parts are similar to the functions of some cell parts	Complex	Functional
Fire analogy	A fire consumes oxygen and fuel and produces carbon dioxide, water and energy; just as a mitochondrion obtains energy from glucose using oxygen, with the same waste products	Complex	Functional

Analogy	Mapped elements	Complexity	Purpose/function
River delta analogy	A river branches into many smaller branches on its way to the sea, just as blood vessels branch into smaller vessels after leaving the heart	Simple	Visual/ geometric
Water pipes analogy	Branching water pipes in a city bring water to houses, just as blood reaches cells through vessels	Simple	Functional and visual/ geometric
Grape analogy	The arrangement of grapes and their stems is similar to the arrangement of alveoli and bronchial tubes in the lungs	Simple	Visual/ geometric

The analogies described above are intended to assist students in constructing content pieces that are of critical importance in understanding the "story" of cellular respiration. Over the three years of our classroom curriculum trials, we were able to evaluate the analogies we used for their effectiveness in promoting learning by students. We also had the opportunity to observe various approaches to using analogies. We revised the curriculum, teacher training, and the teachers' and student manuals each year after evaluating our observations, assessing student learning gains, and receiving feedback from teachers and students. In addition, our cooperating teachers each used analogies slightly differently. In this chapter we summarize some of the things we learned from this formative evaluation process and from our observations of varying methods of teaching with analogies.

12.5 Analogies and Mental Models

Mental models are simplified representations of natural processes that are used both in learning and in the application of what has been learned. There is a growing recognition that mental models are reasoning and learning tools for both scientists and science students (Nunez-Oviedo & Clement, 2003). Mental models help us to make predictions and develop causal explanations (Gentner & Forbus, 1996). Hatano and Inagaki (1988) found evidence that young children build models through the use of analogies, using what they know about their own needs and tendencies to make predictions about animals and plants.

In our own investigations, we found evidence that analogies contributed critical portions of students' mental models. We were able to track the change in students' models by assessing students for understanding of key model elements before and after teaching with an analogy. These "pre/post" assessments showed that the percent of students drawing cells arranged closely together rather than widely separated increased after the use of the ear of corn analogy (Table 12.1, Table 12.2). Gains were even greater in knowledge about the circulatory system. After a series of analogies and discussion of these analogies, students' drawings of vessels showed that the number of drawings in which vessels are branched more than doubled. Similar increases were seen in the understanding that branching progresses from large to smaller vessels. A third critical understanding, that blood returns to the heart via vessels rather than going on a "one-way trip," was shown by nearly twice the number of students post-analogy use. (Table 12.2). This suggests a progression from a simple and incorrect model to a model that included several features important to an understanding of the structure and function of the circulatory system. These findings also suggest that analogies can be used successfully by middle-school aged children.

Table 12.2 Seventh-graders' understandings of cell arrangement and circulatory system structure before and after presentation of analogies

Concept	Pre % correct	Post % correct	Increase from pre→ post (% increase)
Cells are arranged closely	52.4	72.5	20.1 (38.3)
Vessels branch	41.3	100	58.7 (142.1)
Branching is from large to small vessels	32.6	78.2	45.6 (139.9)
Blood returns to the heart	52.2	90.0	38.7 (74.1)

12.6 Maximizing the Learning Benefits of Analogies

Both Bulgren et al., (2000) and Glynn (1991) have developed methods of teaching with analogies that are structured and predictable. Bulgren et al. presented analogies in table form, while Glynn used his six-step "Teaching With Analogies" format, in which the analogy's purpose, target, base, and mapping are introduced in successive steps. In both methods, teachers specify precisely both what does, and what does not map or transfer from the base to the target. This structure is used to help students gain the benefits of analogies, and also to help prevent the "unintended consequence" of analogy use, analogy-induced misunderstandings.

Over the three years in which we conducted trials of the curriculum, we had the opportunity to observe and assess a number of different methods of teaching with analogies. As we revised the curriculum over the years, we and the teachers we worked with, experimented with more- and less-structured ways of helping students access analogies. Our experiences suggest both that there is a need to make analogy use structured and predictable, and that it is important for teachers to be aware that students can misunderstand analogies. We found gains in understanding when analogies were used, but we also found that there were times when students developed misunderstandings that we can attribute directly to our ways of using analogies. These misunderstandings were surprising and sometimes amusing. Students encountering the ear of corn analogy predicted, as was our intention, that cells were closely arranged. Some, unfortunately, also suggested that cells would be hard, colorful, or attached to something like a cob at the base. During presentation of the analogy in which the "burning" of glucose in the mitochondria was compared to the process of combustion in a fire, the teacher carefully detailed the inputs and products that are common to both fires and cellular respiration – fuel, oxygen, water, energy, and carbon dioxide. A student called out, excitedly, referring to the cell: "I know why they need the water! To put out the fire!" This student seems to have inferred that there was an actual fire burning in each cell. We call these two kinds of errors "overmapping" and "mismapping." An "overmap" is the transfer elements of the base that are not intended as parts of the target, as we found with some initial trials of the "ear of corn" analogy. In the complex and unfamiliar fire analogy, we found cases of mismapping, or transfer of base elements in a way other than intended. Glynn (1991), Duit, (1991) and Harrison (2001) have all reported student confusion as a result of analogy use.

In our curriculum revisions, we, therefore looked for ways to both simplify our analogies and structure the process of using them. For example, we simplified the ear of corn analogy so that students look only at the way cells are arranged. We simplified the fire analogy so that students would not be expected to know all the products of the reactions, and instead emphasized the more fundamental notion that in both a fire and the mitochondria, a "fuel" is broken apart to release energy. We structured the process of presenting analogies by giving teachers and students guidance as to exactly which analogy features to attend to, and explicitly calling their attention to features that do not transfer.

We offer the following guidelines for working with analogies, especially at the pre-secondary level. These guidelines are outgrowths of Glynn's "Teaching With Analogies" model with modifications that are based on our observations of teachers' use of analogies in the classroom. These guidelines should allow teachers to be flexible in presentation method while still ensuring maximal student success with analogies.

- *Before beginning, call attention to the fact that the learning tool that will be used is an analogy.* Ask students to recall the purpose of analogies as a means of learning in science. We found that some students had trouble keeping track of the word analogy, even after a number of uses of the word in their classroom. Having students state out loud or write both the definition and purpose of analogies before use serves a metacognitive function in helping students know what is about to happen in their class.
- *Call attention to the purpose of each analogy before beginning the analogy.* Some analogies help students understand how things work or function. Others help students understand how things are arranged or what they look like. Our school analogy, for example, helps students learn that the cell is a place that is divided into smaller places that each have functions. Our ear of corn analogy, on the other hand, helps students get a visual model of cell arrangement and to see cells as arranged closely rather than distributed loosely. We suggest that specifying whether the goal of an analogy is to provide a visual or functional model, or both, is important. In addition, we suggest that teachers call attention to the specific learning goal of each analogy. For example, in the ear of corn analogy, we directed students to look only at how the kernels of corn were arranged, not at color or any other feature of the kernels.
- *Have students spend time familiarizing themselves with the mappable parts of the base.* We gave students a real ear of corn to look at when we used that analogy. We had students look at a graphic of water pipes leading to houses when we used the water pipes analogy to explain the branching nature of vessels. Adding visuals makes analogy use more

engaging and more visual than it would be if the analogy were only explained by the teacher or read by students. In addition, familiarizing students with the base before proceeding ensures that learners are on a "level playing field" and helps minimize differences among learners that are due to culture and experience.

- *Keep the analogy and its discussion as simple as possible.* Simple analogies are more easily understood than complex analogies. We generally found fewer errors in analogies that only had one or a few "mappable" elements than in analogies that had many.
- *Use drawings, diagrams, and tables to show which elements correspond in the base and the target.* In the school analogy, for example, we used a table in which parts of the school were placed in the left-hand column and the parts of the cell that had similar functions were placed in the right-hand column. At pre-high school levels, however, tables may need to be introduced gradually, particularly if students are unfamiliar with their use.
- *Call attention to the parts of the analogy that do not map.* After use of the ear of corn analogy, students were asked questions such as: "Do you think other features of the corn also apply to cells? Do you think they are hard? Feel the cells of your arm. Are they hard? Do you think they are colorful, the way the kernels are? If your cells were the colors of the corn kernels, what would you look like?" Students laughed and appeared to think that these statements were obvious. Nonetheless, our previous assessments had shown that some students make such errors when these distinctions are not drawn.
- *Make analogy use as student-active as possible.* We suggest that teachers try to make maximum possible use of student voices by asking questions rather than making statements. Students can often contribute descriptions of the base and target and predict elements that correspond. Students who go into "passive" mode may easily get confused during the classroom use of analogies.
- *Assess understanding.* Teachers should assess understanding of the model elements taught by the analogy. This allows both teachers and students to monitor the process of learning with analogies in the classroom. Our observations and assessments suggest that student confusion was a danger with our more complex analogies. This confusion was generally not apparent during class discussions. Individual or small-group assessments may, therefore, be useful. Assessments can be as simple as asking students to draw or diagram what they have learned from an analogy. Such assessments need not be time-consuming, and can be used to reinforce what has been taught.

These procedures were observed in use by participating teachers who used analogies successfully, with the exception that teachers did not tend to assess students to determine to how many had confusions or misperceptions about the analogy. Our observations suggest that students who are more vocal in class discussions tend to have a greater degree of understanding and engagement than some other students. We therefore stress the importance of assessing all students' ideas to determine which model elements have been learned after analogy use. This assessment can lead to needed modifications in the methods used in presenting and discussing analogies.

12.7 What Factors Should be Taken into Account in Designing a Lesson that Uses an Analogy?

We found that students' understanding of our analogies varied from analogy to analogy. This variation was observed despite the fact that our teachers were experienced in the use of analogies and used similar presentation methods for all analogies. We must therefore ask whether some analogies are harder to comprehend than others, and, if so, what factors should be taken into account in designing a lesson that uses an analogy. It may be that there are features of analogies that can help us determine when we need to take extra care when we use them. We will begin by reviewing the literature and our own earlier work on this subject. The following features distinguish some analogies from others, and may be associated with analogy effectiveness.

Near vs. far: Gentner (1989) identified analogies as being either "near" or "far." Near analogies are those in which the base and the target share a number of superficial similarities, and far analogies share few similarities. For example, our school/cell analogy is far because a school and a cell share few surface similarities, and are similar only in the relationships among the elements of each. Gentner's prediction was that far analogies would be more effective than near because they would increase learners' level of awareness. On the other hand, Newby et al. (1995) predicted that the "distance" between base and target domains would decrease the ease with which an analogy was understood by students.

Familiar vs. unfamiliar: Do analogies need to be familiar to students to be useful? Goswami (1992) suggested that analogies that are already familiar to students are superior to those that are not. Stavy (1991), however, suggested that an unfamiliar analogy can be made accessible to students through explanation or experience. Harrison (2001) describes a chemistry analogy about a high school dance, and suggests that the analogy's interest for students increases their motivation and attention.

Visual/structural vs. functional: Our data lend some support to the idea that it is important to be aware that some analogies can help students understand what something looks like, while others help learners understand what something does or how it works. Still other analogies convey both structural and functional information. Curtis and Riegeluth (1987) suggested that analogies that convey both structural and functional information are understood and retained more effectively than those that convey only one type of information.

Simple vs. complex: In some analogies, a number of features of the base map or correspond to the target. In others, only a few correspond. Analogies in which students are guided to map only one or a few features are simpler than those in which they are asked to map a number of features. In other work (Else et al., 2003) we suggested that simpler analogies are more easily understood than complex analogies.

Table 12.3 Elements mapped correctly from fire to the mitochondria in a post- assessment of the fire analogy

	Oxygen	Fuel (food, glucose wood or other fuel)	Energy output	Carbon dioxide	Water
% students mapping correctly	40.3	74.6	65.7	47.8	32.8

We were able to identify features of analogies that may be used to characterize them, and find large differences in students' reproduction of target model elements following use of the analogies. We hypothesize that some analogy features may be associated with the degree of difficulty in comprehending an analogy. For example, in the *complex* and *unfamiliar* fire analogy students were asked to compare the inputs and products of a fire – fuel, oxygen, energy, carbon dioxide, and water – to the similar inputs and outputs of the mitochondria. We found that students mapped some of the elements correctly but had trouble identifying others, or confused inputs and products (Table 12.3). Fewer than half of the students were able to identify water as a product of cellular respiration, for example. We suggest that this is a difficult analogy at least in part because of features such as its unfamiliarity and complexity.

On the other hand, use of the *simple* and *familiar* corn analogy resulted a dramatically higher post score after a relatively brief experience with the analogy. The number of students who drew cells as widely separated rather than contiguous was reduced by more than half following use of that analogy (Table 12.4). Readers may be surprised by the fact that not all students responded to even this simple analogy. This may be partly due to the fact that this was the first explicit analogy that students had been asked to comprehend. But it also signals that the meaning of an analogy, no matter how obvious to us, may be very different for a young student. We might also speculate that the tendency to see cells as separated is a widespread and powerful misconception, introduced perhaps by students' prior experiences with single-celled organisms.

Table 12.4 Students' ideas about the arrangement of cells in the body pre- and post- instruction. Instructional methods included the ear of corn analogy and small- and large-group discussion of the analogy

Concept	Pre %	Post %	Change
Cells arranged closely, with little or no space between them	52.4	72.5	+20.1
Cells separated	33.3	14.5	–18.8
Drawing unclear	14.3	13.1	–1.2

We also found that students occasionally confused the functional and visual/structural features of analogies. For example, some students drew analogies for cell organelle function as their visual representations of the organelles. When questioned, these students seemed genuinely confused, and had clearly misunderstood the purpose of the analogy. While some of our results can be attributed to the fact that our learners were young and unfamiliar with the instructional use of analogies, we also think that these results underscore the need to assess student understanding following use of an analogy. Our formative evaluation results therefore suggest that teachers need to be alert to the potential for student confusion, particularly when complex and unfamiliar analogies are employed. We hypothesize that these factors make an analogy more difficult to comprehend and that such an analogy should not be used if these factors make teaching an analogy prohibitively difficult for a given age group.

12.8 When are Analogies the "Right Tool?"

We consider analogies to be useful to students who are learning about the human body because they help students build visualizable mental models that are transitions to "target" models – models that are like scientists' understandings. As shown in Table 12.1, we believe many also serve to provide the foundation for new visual imagery. We have observed that students tend to find analogies engaging and approach them actively, suggesting that analogies may serve motivational functions. In addition, because analogies must be thought through to be understood, we suggest that analogies encourage active student construction even in students who are not accustomed to thinking actively in school science.

We recognize that, to be understood by students, analogies, especially complex ones, require highly-structured and intensive processing. As Newby et al., (1995) have noted, as a consequence, analogies are "effective but not efficient." We therefore consider analogies to be "expensive" teaching tools, and reserve them for concepts that are not accessible to students through demonstrations or experience. The use of analogies is also reserved for important or abstract concepts and concepts that are prerequisites to further learning.

The ear of corn analogy is a case in point. Research suggested that some students see body cells as having no particular arrangement, as being loosely-packed and not contiguous. We deem the understanding that cells are, on the contrary, arranged contiguously, with little intercellular space, to be quite important. Students will, later in the curriculum, be expected to be able to develop understandings of oxygen and carbon dioxide transfer between cells and the blood in capillaries. In order to construct the correct model of the capillary's proximity to the cell, students must have a model in which cells are contiguous.

We use analogies to introduce the target concepts in Table 12.1 – cell arrangement, cell parts, mitochondrial inputs and outputs, blood vessel and lung structure – because they can not easily be built by students themselves. In contrast, students are rather easily able to construct some of the curriculum's other models. For example, students developing understandings of the digestive system quickly understood that we have two "tubes" in our throats – the esophagus and the trachea. With some prompting by teachers, students are able to access such experiences as choking, burping, and swallowing, which students can use as "clues" when they try to infer their own internal structure. Most students can comprehend, with teacher support, the idea that if they did not have two tubes, they would be likely to get air in their stomachs and food in their lungs.

Lastly, we have chosen to use analogies as a learning tool when effective analogies are available. As discussed above, one criterion of effectiveness is familiarity and accessibility of key relationships in the base, or the ability to fill these in. The ear of corn analogy, for example, is made accessible to students by giving them an actual ear of corn to look at, hold, and draw. The fire analogy to mitochondrial respiration is explored by lighting a candle in a jar, then covering the jar until the flame dies. This demonstrates to students that oxygen is needed for combustion, a fact that may be unfamiliar to some students.

12.9 Summary and Conclusions

Analogies are one of a number of tools used in the *Energy and the Human Body* curriculum. In this paper we have we reflected on patterns in our observations of strengths and weaknesses in analogy use in our classroom trials. We have used these reflections to develop hypotheses about techniques for using analogies in instruction. These hypotheses should be subjected to evaluation and improvement in further research. We have come to use analogies in areas of the curriculum where students cannot readily build models themselves through inference, invention, or experience. We believe that the processing students use in examining and understanding the analogies included in the curriculum is active, in that students are asked to examine and reflect on each analogy. We acknowledge that much of the difficulty in comprehending some analogies may have been a function of the fact that our students were both young and inexperienced with analogies. We would also note that some analogies used in our curriculum were spectacularly successful, generating active, engaged discussion among teachers and students and being recalled by students months after use on the classroom. We give preference, however, to starting from students' own models when they are available, and use analogies where preliminary research indicates that a concept cannot be readily developed by students. As an alternative to analogies, students and teachers are able to marshal familiar concepts and combine them in a new model, with either teacher or students taking the lead.

References

Brown, D. E., & Clement, J. (1989). Overcoming misconceptions via analogical reasoning: abstract transfer versus explanatory model construction. *Instructional Science 18*, 237–261.

Bulgren, J. A. Deshler, D. D. Schumaker, J. B. et al (2000). The use and effectiveness of analogical instruction in diverse secondary content classrooms. *Journal of Educational Psychology* 92, 426–441.

Clement, J. (1993) Using bridging analogies and anchoring intuitions to deal with students' preconceptions in physics. *Journal of Research in Science Teaching* 30,1241–1257.

Curtis, R. V., & Reigeluth, C. M. (1984). The use of analogies in written text. *Instructional Science, 13*, 99–117.

Dagher, Z. (1994) Does the use of analogies contribute to conceptual change? *Science Education 78*, 601–614.

Duit, R. (1991). On the role of analogies and metaphors in learning science. *Science Education 75*, 649–672.

Else, M. J. (2003) Should different types of analogies be treated differently in instruction? Observations from a middle-school life science curriculum. *Proceedings of the National Association for Research in Science Teaching*, Philadelphia, PA

Gentner, D. (1989). Mechanisms of analogical learning. In S. Vosniadou, A. Ortony, (Eds.), *Similarity and analogical reasoning*. (pp. 199–241). London: Cambridge University Press.

Gentner, D., & Forbus, K. D. (1996). Analogy, mental models, and conceptual change. http://www.qrg.northwestern.edu/projects/ONR-SM/analogy.htm

Glynn, S. M. (1991). Explaining science concepts: A teaching-with-analogies model. In S. Glynn, R. Yeany, B. Britton, (Eds.), *The psychology of learning science*. (pp 219–240). Hillsdale, NJ: Erlbaum.

Glynn, S. M., & Takahashi, T. (1998). Learning from analogy-enhanced science text. *Journal of Research in Science Teaching, 35*, 1129–1149.

Goswami, U. (1992). *Analogical reasoning in children*. Hillsdale, New Jersey: Lawrence Erlbaum Associates.

Harrison, A. G. (2001). Thinking and working scientifically: The role of analogical and mental models. Paper presented at the annual meeting of the Australian Association for Research in Education, Fremantle.

Hatano, G., & Inagaki, K. (1988). Constrained person analogy in young children's biological inference. Annual Meetings of the American Educational Research Association, New Orleans.

Johsua, S., & Dupin, J. J. (1987). Taking into account strategy conceptions in instructional strategy: An example in physics. *Cognition and Instruction, 4*, 117–135.

Newby, T. J., Ertmer, P. A., & Stepich, D. A. (1995). Instructional analogies and the learning of concepts. *Education Technical Research and Development, 43*, 1042–1629.

Núñez-Oviedo, M. C., & Clement, J. (2003). Model competition: A strategy based on model based teaching and learning theory. Paper presented at the National Association for Research in Science Teaching Conference, Philadelphia, PA.

Rea-Ramirez, M. A. (1998). Models of conceptual understanding in human respiration and strategies for instruction. In D. Thesis, (Ed.), University of Massachusetts, Amherst.

Rea-Ramirez, M. A., Nunez-Oviedo, M., & Else, M. J. (2002). Energy in the human body teachers' manual. University of Massachusetts, Amherst.

Stavy, R. 1991. Children's ideas about matter. *School Science and Mathematics*, 91, 240–244.

Spiro, R. J., Feltovich, P. J., Coulson, R. J., & Anderson, D. K. (1989). Multiple analogies for complex concepts: Antidotes for analogy-induced misconception in advanced knowledge acquisition. In S. Vosniadou, A. Ortony (Eds.), *Similarity and analogical reasoning*. (pp. 498–531). Cambridge: Cambridge University Press.

Chapter 13
Model Based Reasoning Among Inner City Middle School Students

Mary Anne Rea-Ramirez

Western Governors University

Maria Cecilia Núñez-Oviedo

Universidad de Conception

13.1 Introduction

There has been an increasing recognition of science as a social process that involves conjecture and argumentation (Newton, Driver, & Osborne, 1999). This recognition implies that scientific ideas may not be easily discovered by the students through empirical practices only. Rather, learning science "involves being initiated into the ideas and practices of the scientific community and making these ideas and practices meaningful at an individual level" (Driver, Asoko, Leach, Mortimer, & Scott, 1994, p. 6). In this initiation process, the teacher plays a key role by providing experiences and helping the students to build the scientific conventions. These activities may take the form of a dialogue or conversation through which the teacher and the students exchange ideas to produce conceptual change. In other words, the teacher, instead of "pouring knowledge into the students' head", becomes a partner that guides the student in the process of co-constructing knowledge (Rea-Ramirez, 1998). This co-construction can also be seen as a process of shared reasoning (Resnick, Salmon, Zeitz, Wathen, & Holowchak, 1993). All too often, however, inner city school students find themselves missing out on rich classroom interactions, alternative educational experiences, and positive social interactions due to a variety of educational conditions. This may include issues with lack of funding, poor teaching models, or lack of belief that students can learn.

13.2 Background

Inner city students participated in research trials on a curricular product that used an approach to teaching middle school biology where students play a much more active role in model construction than is the norm. "Energy and the Human Body" curriculum deals with pulmonary and cellular respiration, circulation, and digestion at the 7th – grade level (Rea-Ramirez, Nunez-Oviedo, Clement & Else, 2005). The strategy is one of student-teacher co-construction that elicits student-generated model elements as well as some that are introduced by the teacher (Rea-Ramirez, 1998; Nunez-Oviedo, et al., 2002). This means that the students are active participants in the inquiry processes of model building, but the activity is guided and fostered by the teacher to insure meeting the targeted content goals. Based on earlier research into how students construct understanding in this area, and based on model based reasoning theory, this curriculum has been tested over five years in a number of public schools in the Northeast and Northwest United States (Nunez-Oviedo, 2004; Rea-Ramirez & Else, 2003).

In light of the above findings, and having achieved significant gains in other trials (Rea-Ramirez, 1998; Nunez-Oviedo, 2004), we wondered whether this model of teaching and learning could be successful with inner city students as well as those from more affluent university town schools or poor rural schools. With this in mind we proposed to explore ways of assessing gains in conceptual understanding in a model based instructional setting for inner city life science students. Several aspects of "conceptual understanding" were examined. (1) The ability of students from inner city schools to develop integrated conceptual models of biological systems, when the teacher acts as facilitator in the co-construction of mental models; and (2) The students' ability to reason about new situations using their constructed model.

Seventh grade students from a largely Hispanic (70%) inner city middle school were recruited to participate in an after school program that taught the curriculum. Students in this city were chosen for inclusion in the trial because they come from a school district that faces similar problems in inner city education seen in larger cities across the nation. While a city of only 40,000 inhabitants, it has the highest percentage of children living in single parent households in the state as well as the highest percent of children in public schools receiving public assistance. In addition it has the highest incidence of reported child abuse and neglect, the large number of foster children and children living in homeless shelters, and the highest percentage in the state of children ages 5–17 living in poverty (51%). The median household income is $22,858 considerably lower than the county average and much lower than the national median of $42,504 for whites and $30,735 for Hispanics. The local school district ranked among the three lowest in the state on statewide academic testing (MACS).

Students in the study were self-selected from a general notice and call for participants sent out by the school to all seventh graders in the school district (approximately 600 students). After attrition, 18 students completed the entire program. The final group consisted of 8 boys and 10 girls, with a diversity of ability levels as measured by science grades in school at the time of their participation (A- to A = 4; B- to B = 2; C- to C+ = 7; D = 3; F = 2). The ethnic makeup of the group included 13 Hispanic, 2 Caucasian, 2 African-American, and 1 Asian, consistent with the approximate percentages of ethnic groups in the school system. Two of the Hispanic students were listed as limited English proficiency by the school district. However, at least three other students displayed very low ability in reading and writing in English and Spanish on classroom activities and assessments. The distribution of Reading/ELA grades for students was comparable to their science grades (A=4; B= 2; C=7; D=3; F=2). These Reading/ELA scores were not confined to Hispanic students, however, with three of the lowest grades belonging to students of other ethnicity. The class was divided into five small groups. There was an attempt to have each group or table with gender, ability, and ethnically mixed organization. When necessary, the teacher moved the students from one group to the other throughout the sessions. The teacher (Rea-Ramirez) was the lead author of the curriculum and had conducted previous research on preconceptions and tutoring strategies in this area (Rea-Ramirez, 1998). Thus, we sought to investigate whether comparable gains could be achieved in the inner city setting, and/or whether there would still be major adjustments and new factors in the inner city setting. Thirty hours of instruction were conducted in total with the students.

Prior to and after instruction students were evaluated using multiple choice and open-ended tests. Student discussion segments were investigated using transcripts from videotape with subsequent analysis and coding to illustrate the types of model based reasoning the students were able to engage in during instruction. The pre/post test consisted of a 35-question multiple-choice test that covered a selection of key topics from the curriculum. Questions were clustered around understanding of the target concepts and represented more concrete, structural knowledge rather than application. In addition, an open-ended assessment prompt was also given to determine the students' ability to reason with their model and to transfer what they had learned to a new situation and to show whether they had developed an integrated model of respiration that included integration among body systems at an organismic and cellular level. The question used a brief story about a hiker who attempts to hike up a mountain without first eating breakfast to stimulate student discussion of the integrated model of respiration. He exhibits symptoms of increased heart rate, increased deep breathing, and generalized weakness. Drawings and written responses on the open-ended question were scored with a rubric that addressed each of the target concepts.

To answer the hiker question with a maximum possible score, the student had to have developed a model of respiration where, if they added or removed an element, they could run the model and see where in the system it would break down and what symptoms this would exhibit in the body such as shortness of breath or increased heart rate.

Each student's pre and post tests were analyzed for whether they:

- Showed a model that was static or dynamic (see key on Table 13.4 for definition)
- Showed causal reasoning, and whether they
- Showed knowledge of structure and function and integration of the systems and cell according to the rubric that reflects the target concepts and optimal integrated model.

In addition, students' models were then compared to the Human Respiration Framework (Figure 13.1) graphically to represent the level of integration and understanding. In order to achieve a certain level, students were required to show understanding of at least 80% of the components in individual sections of each level. Therefore, it was possible for students to be a level II in area A that is the pulmonary system and level IV in area D, the digestive system. The optimal target is Level III A, B, D and IV C. However, in previous research, students who have scored level II A, B, D and IV C have shown dynamic, causal, and integrated models with the ability to use this model to reason about new situations.

13.3 Evidence of Conceptual Learning

In addition to determining whether significant pre post gains existed within this experimental group, researchers were also interested in how these inner city students performed when compared to students from other school districts. This included a small university town middle school (School 1), a rural, predominately low-income middle school (School 2), and a middle class urban school (School 3). Table 13.1 compares the pre-post differences on the multiple choice portion of the pre-post test among schools, with School 4, indicating the low income, largely Hispanic, inner city school. While students' pre test scores were lower than the other schools, pre-post differences showed gains comparable to all other schools.

Table 13.1 Comparison of multiple choice scores for all schools

School	% Correct pre-test	% Correct post-test	Pre/post diff	df	Significance
School 1	39.54	63.73	24.28	68	$p<.001$
School 2	31.34	59.57	28.23	66	$p<.001$
School 3	41.17	62.40	21.23	43	$p<.001$
School 4	28.89	55.56	26.67	17	$p<.001$

On the open-ended application problem there was also an overall significant pre-post gain ($p<.001$) (Table 13.2). When compared to the other schools also studied at the same time, students in the inner city group scored higher overall in the open-ended application problem.

13.3.1 The Human Respiration Framework

The Human Respiration Framework (Fig. 13.1) was developed in an attempt to visualize student levels of understanding of respiration both at a structural and functional level. The framework evolved out of studies on

Table 13.2 Comparison of open ended scores for all schools

School	Pre-test mean	Post-test mean	Diff mean	Sd	Significance
School 1	1.99	9.39	7.40	4.4	<.001
School 2	2.13	6.74	4.62	4.05	<.001
School 3	2.45	8.41	5.96	2.2	<.001
School 4	1.24	10.34	9.52	4.61	<.001

student constructed understanding in which the students' knowledge was graphically organized individually and then compared among students. This led to the present graphic that incorporates naive understandings, beginning understanding, and more advanced understanding. That is, many students' initial understanding of organismic and cellular respiration was very naive, at the I level on the Human Respiration Framework (HRF). However, some students were able to describe respiration at a higher level such as II or III. Often these students would describe some structures and functions at one level and others at another. Therefore, it became evident that not only were levels of understanding present but that within levels it was necessary to identify systems that interact (in Fig. 13.1 these are designated as A, B, C, and D.)

An interesting finding emerged early in the research indicating that a few students were able to develop and explain an integrated, complex, and dynamic model of respiration down to the cellular level that did not include an advanced level of knowledge about individual structures. That is, when students understanding was placed on the HRF, they might be at a Level II or III in Sections A, B, and D but a Level IV in Section C. The HRF became an important tool for analyzing, understanding and describing in depth students' knowledge of respiration and how integrated, dynamic, and complex structurally this model was. The HRF was used in this study to document students' mental models depicted on the pre and post tests. Using a rubric, we were able to accurate determine students' Level and Section understanding and then to compare these pre and post.

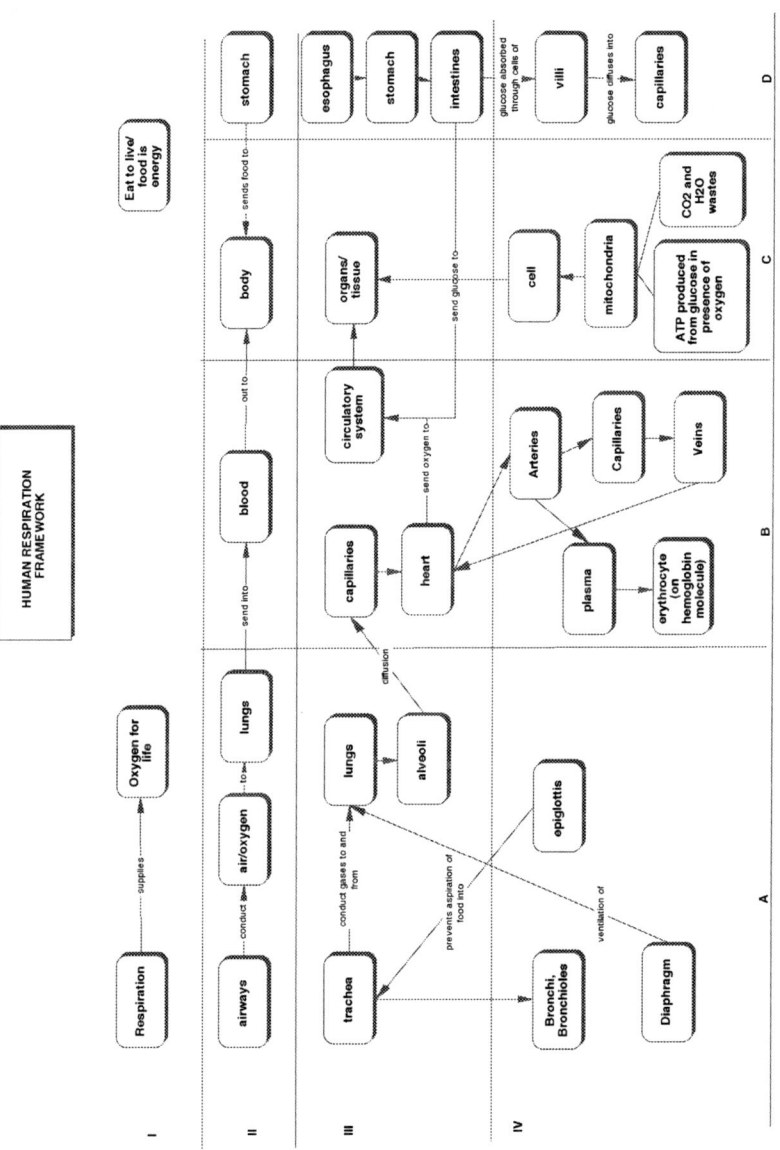

Fig. 13.1 Human Respiration Framework showing the structure and function of each system, the integration between systems and the minimal learning pathway necessary to show conceptual understanding

13.3.2 Student Understanding on Open-ended Question

13.3.2.1 Pre Test

Students' responses on the open-ended question were further analyzed for evidence of knowledge of structure and function, increase or decrease in dynamics and integration, and inclusion of causal chains, with subsequent placement of students on the Human Respiration Framework to show change pre to post instruction. Table 13.3 shows students scores in the individual categories and overall Human Respiration Framework levels on the pretest. Most students (13) were at a Level I for all sections, while 4 others Only one of these, Student 8, showed integration of the three systems, but no integration between the systems and the cell. This student's model showed minimal evidence of dynamics (describes food being digested and absorbed into the blood where it flows and gives energy). At the same time, Student 8's explanation also contained a number of alternative conceptions. No students were able to describe a causal chain. Most attempts to give a cause-effect for the symptoms simply stated that the hiker was out of breath, tired, or because he lacks energy. There was no evidence of a model of respiration or of using a model to explain symptoms. Only four students included any structure in their drawing or description and only two included function, although in both of these cases function included multiple alternative conceptions. These included, "Food is digested and turned into a gas", "Energy is floating in the blood", "The heart is cleaning the blood." Structures that were primarily identified by the four students were the heart, lungs, and stomach. Only two of these students actually made a drawing that included some or all of these structures.

Table 13.3 Analysis of pre instruction student mental models

Key: **static** = student does not show motion or flow through the systems or between systems and cellular level or within cell
Dynamic = student shows evidence of flow through the systems and/or between systems and/or cellular level
CAUSAL = student suggested more causal relationships and a higher level of logical thinking than expected to meet the minimal integrated target model
causal = student suggested minimal causal relationships using model to explain
non-causal = student did not show evidence of any causal relationships in model or causal reasoning does not use model
S = average to above average structures; **s** = minimal structure
F = average to above average function **f** = minimal function
Blank = no structures/functions noted

Student	Static/ dynamic	Causal non-causal	Structural/ functional	Integrated: organismic and cellular between systems	Initial level (Human respiration framework)
Student 1	static	non-causal		no integration	I
Student 2	static	non-causal		no integration	I
Student 3	static	non-causal		no integration	I
Student 4	static	non-causal		no integration	I
Student 5	static	non-causal		no integration	I
Student 6	static	non-causal		no integration	I
Student 7	static	non-causal		no integration	I
Student 8	minimal dynamic	non-causal	s/f	some integration of systems but not to cell	IIA,B,C
Student 9	static	non-causal		no integration	I
Student 10	static	non-causal	s/f	no integration	I,IIB
Student 11	static	non-causal		no integration	I,IID
Student 12	static	non-causal		no integration	I
Student 13	static	non-causal	s	no integration	IIB,D
Student 14	static	non-causal		no integration	I
Student 15	static	non-causal		no integration	I
Student 16	static	non-causal	s	no integration	IID
Student 17	static	non-causal		no integration	I
Student 18	static	non-causal		no integration	I

13.3.2.2 Post Test

Analysis of post data indicated that many students increased in their understanding of respiration, particularly structure and function, dynamics, causality, and integration (Table 13.4). However, while it was evident from the inner city students' answers that they had developed complex models of respiration, some had difficulty translating this into explanations of everyday phenomena. In many instances where students attempted to talk about the symptoms the hiker showed, they reverted to naïve answers rather than to answers that incorporated the complex model they had just constructed. That is, students often could speak about respiration in detail, with particular emphasis on function down to the mitochondrial level, but often did not relate it back to why the hiker's heart beat increased or his breathing became deeper and faster. Application and mental runnability of a model is considered an extremely important skill (Clement, 2003). This allows students to make connections to real life and to apply learning to new situations.

It was suggested from the open-responses (both drawn and written) and post scores on the Human Respiration Framework that students who had an integrated understanding of organismic and cellular respiration as opposed to an isolated system knowledge, were better able to discuss the situation of the hiker in terms of what was happening within the body down to the cellular level. That is, students would have to include all sections A, pulmonary, B, circulatory, C, body, and D, digestive, and score at least a Level II on each of the sections A, B, and D, and score a level IV on section C to show integration across systems and integration between the systems and the cell. However, students with the isolated system knowledge were able to discuss the structure and function of the system but showed little if any ability to discuss what was happening in that system in relation to the symptoms the hiker experienced. Ten students attained Human Respiration Framework levels at or above the target (II A, B, D, IV C), while another 4 were at a level of integrated understanding for systems but did not include integration between systems and the cell.

In the next section, we will examine through student drawings and reasoning sequences evidence of model based reasoning and understanding at target concepts.

Table 13.4 Analysis of post instruction student mental models

Student	Static/ dynamic	Causal non-causal	Structure/ function	Integrated organismic and cellular and/or between systems	Final level (Human respiration framework)	Open ended test pre-post difference
Student 1	dynamic	causal	s/f	no integration	I, III D	4
Student 2	static	non-causal	s/f	some integration between systems and cell	II A,B, IIIC	12.2
Student 3	dynamic	non-causal	s	minimal integration	II A, III B, C, D	6.2
Student 4	dynamic	causal	s/f	Integration between digestive system and cell only	IIB, IV C, III D	9.6
Student 5	dynamic	causal	s/f	Integration	III A, IV B,C,D	15.8
Student 6	dynamic	non-causal	s/f	Integration	II A, IV B,C,D	12.6
Student 7	static	non-causal	s	no integration	II A, B, C, D	4.8
Student 8	dynamic	causal	s/f	Integration	III A, B, IV C, D	19.8
Student 9	dynamic	causal	s	minimal integration	II A, B, IV C, I D	8.8

Student 10	dynamic	causal	s/f	minimal integration	II A, IV C, D	8.8
Student 11	dynamic	non-causal	s	minimal integration	II A, III B, C, D	4.4
Student 12	dynamic	causal	s/f	integration	III A, B, IV C	12.6
Student 13	dynamic	causal	s/f	integration	III A, B, IV C, III D	15.8
Student 14	dynamic	non-causal	s/f	no integration	II B, III C, D	8.4
Student 15	dynamic	causal	s/f	integration	IV C, III D	7.2
Student 16	dynamic	causal	s/f	integration	IV A, IIIB, IV C, D	16.6
Student 17	dynamic	causal	s/f	integration	III A, B, IV C, D	16.2
Student 18	static	non-causal	s	no integration	I	3.4

13.3.2.3 Analysis of Students' Reasoning with a Model

Students who scored the highest on the open explanation response showed evidence of using a dynamic, causal model. However, the level of causality varied from minimal to major causal relationships. In only two instances were students able to discuss why symptoms appeared based on what was happening internally, down to the mitochondria level, providing evidence in drawings and words of reasoning with a model. For these students beginning at the symptoms and working "backwards" using the model was indicated by their written discussion and drawings. However, other students were only able to discuss some causal relationships that did not require diagnostic processes or working backwards through their model. They had more difficulty discussing what was happening in the body in relation to the symptoms presented and displayed varying degrees of reasoning with a model. In some cases they were able to begin at the cellular level and reason with a model "forward" following the direction of causes through cell and/or the systems, and integrating systems while not others. However, this progression in most cases did not carry them through to an explanation of the symptoms. It was interesting, then, to wonder what it was that made it possible for some students to attain this level of complexity and others not even though they all appeared to develop dynamic, integrated models.

Most students, while scoring relatively high on the post test, were unable to show reasoning with a model. One example is Student 6 who scored a 0 on the pretest and a 12.6 on the posttest where he drew a picture with all the important target structures, integrating the pulmonary, digestive, and circulatory systems and the cell. In a series of additional drawings, the student detailed the structures and function of the cell, the digestive, pulmonary, and circulatory systems, with the most description given of the digestive system. He then gave a description of the flow between these systems, including an interesting analogy. While he mentioned glucose and oxygen being transported to the cell by the circulatory system, he did not, at this point, discuss where they came from.

> "The *hert [hear]) pumpet [pumped]* out to the *arrays [arteries]*, the *arrays [arteries]*
> *brake [break]* down and become *capulary's [capillaries]*.
> The *capulary's*
> *[capillaries]* become the table *fore* the cell. The cell get the *glucoses* and
> oxygen and give in CO2 and H2O and *thay [they]* go to the *veans [veins]* and back
> to the *hearth [heart]*."

While this discussion shows a dynamic model with flow through the systems to the cell, it breaks down when Student 6 attempts to provide cause-effect relationships. At this point he reverts to naïve ideas about breathing and is unable to use the model he has constructed to explain the symptoms presented by the hiker.

> "His *hearth* is going fast *becose [because]* he is not *birthing [breathing]* well.
> Where he is taking a lot of *are [air] becose [because]* he is not getting *inofe*
> *[enough]* air. The lungs are not getting air. *Thay [they]* gave all *there [their]* air."

As a result, this student was not able successfully to integrate all the elements learned to explain scientifically the symptoms experienced by hiker.

In another example, Student 8 scored a 3.2 on the open-ended pre test, showing minimal integration between the circulatory and digestive systems but no integration between other systems or the cell. Her model was minimally dynamic in that she indicated flow of energy through the system (Fig. 13.2). However, her model included numerous alternative conceptions such as:

> "Food is digested and turned into a gas and the gas is absorbed into the blood
> where it flows and gives energy."

This explanation sees food as a source of energy but does not say that it must be broken down and transported to the cell where is utilized in the production of ATP (high energy molecules) in the mitochondria. It also describes food being broken down or turned into a gas. Interestingly, in her drawing there is not a digestive system or a circulatory system although there is a heart and energy shown flowing in the body giving her an "s" for structure. For function she scored an "f" for attempts to discuss function of

Fig. 13.2 Pre drawing showing lungs, heart, muscles and energy flowing in the body

the digestive system even though this included alternative and naïve conceptions. Placement on the Human Respiration Framework was at the II A, B, C level since her model and descriptions based on other statements shows basic knowledge of air moving into the lungs, into the blood, and circulating in the body (no mention of organ or cellular level), and while she describes the need for digestion, she does not describe the system necessary for this.

On the posttest, however, Student 8 gave a much more complete explanation of energy in the body and completed a series of drawings to support her description. In addition, where she did not relate the actions of rapid, deep breathing or increased heart rate to what was happening in the body on the pre test, she now attempts to use causal reasoning to provide explanations for the symptoms according to what is occurring within systems and the cells.

> "The friend is using energy to hike, but the hiker didn't eat any food so she is not going to get any glucose from the food. Also she is breathing to get oxygen that is also used with the glucose to get energy. She is breathing hard trying to get oxygen for energy. Her heart is beating fast because she is hiking, while all this is happening, and the muscles are running out of energy and they need more energy so the heart is pumping fast, trying to bring oxygen to get energy to the muscle. Oxygen is transported from breathing oxygen into the lungs, then the heart picks up the oxygen by diffusion and the oxygen diffuses into the blood and the heart pumps the blood through the body and the oxygen diffuses into the cells. The glucose is in the food that you eat and it goes into the stomach where the food is broken down, then it goes to the small intestines where it breaks up the food into glucose and the villi diffuses the glucose into capillaries where the glucose is transported through the blood to the cells."

Student 8 has shown a dynamic and integrated model in that there is flow of elements necessary for energy transformation in the cell through the pulmonary, digestive, and circulatory systems. Evidence indicates that knowledge of both structure and function has increased. In addition, her model shows evidence of causal reasoning as she attempts to give concrete explanations of the symptoms relating them to actual actions and deficits occurring at a cellular level. While some of her explanations still appear naïve by scientific standards, they meet the target concept expected at this level.

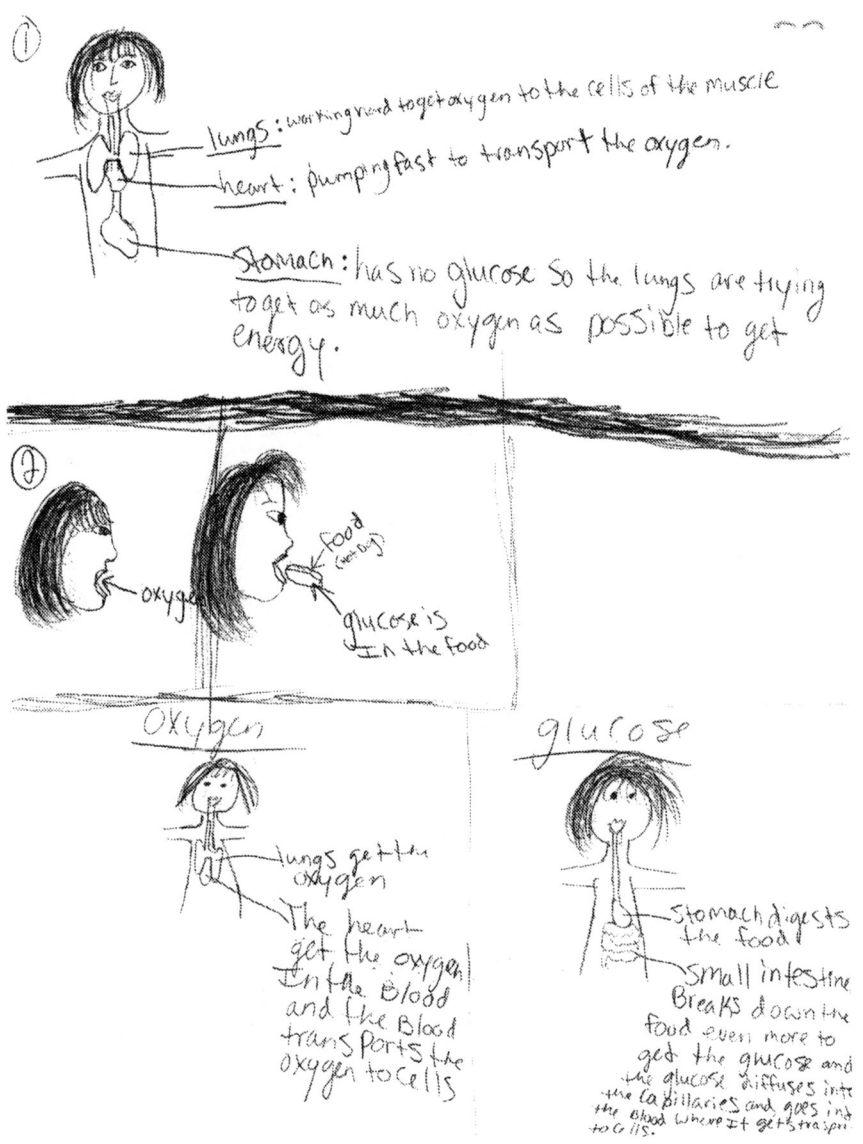

Fig. 13.3 Drawings from Student 8 post test

Another example that illustrates students reasoning with a model is Student 12 who displayed at least a beginning understanding that food was necessary for energy.

> "Since my friend didn't eat she, doesn't have enough energy. She
> breathing so hard, and her heart is beating fast, because I think she dehydrating."

There are no drawings or discussion of systems or cells and no description of function. In addition, her attempts at causal reasoning are at a minimal level with no explanation or causal chain with more than two elements. This gave her a HRF level of I that shows understanding of needing to eat to live or for energy. No connection is made of the need for oxygen.

On the posttest, however, Student 12 shows in diagrams and written discussion, that she has a much more integrated, dynamic, and causal understanding of energy in the human body. A decided difference is noted when compared to Student 8 above. Student 12 discusses oxygen going to the cells to produce ATP independently from the need for glucose in the mitochondria. While Student 12 shows glucose in the drawing with the mitochondria, she does not describe where it comes from except in very general terms, i.e.

> "When you eat your food give you energy. And (ATP) by the mitochondria. So my friend needed to have eaten so she would have had lots of energy to go up the second hill."

Interestingly, her causal reasoning for breathing faster and deeper initially appears to be naïve.

> "The reason why she/he is breathing so hard is because *sense (since)* he/she *dosen't (doesn't)* have as much energy that person is getting tired, and there's less oxygen when *your (you're)* going up the hill."

However, she then goes on to give an explanation that denotes a deeper conceptual understanding.

> "Another reason why that ... look" (she draws Fig. 13.4 below) with the explanation, "What I'm trying to *discribe*

> *[describe]* is there's less oxygen going into your cells, so they reproduce less ATP."

Student 12's use of the phrase "What I'm trying to describe..." after directing us to her drawing, suggests that she is reasoning with her model, following a path through the integrated systems, including the cell. That is: theorizing there is less oxygen from the lungs to the alveoli, to the bloodstream, to the cells, providing less oxygen in the cells for production of ATP.

On the second drawing in Fig. 13.4 she shows, however, only glucose going into the mitochondria, not oxygen, and says that the mitochondria reproduce energy. Since there is no reference to ATP, it is difficult to determine whether she has a full understanding of relationship between both oxygen and glucose in the transformation of energy to ATP in the mitochondria. While structures of the digestive system are missing, there is sufficient drawing of structures in the pulmonary and circulatory systems and some at the cellular level, to meet the target concept. Her model is also considered dynamic because she has traced the oxygen path through the pulmonary and circulatory systems to the cell, also showing integration between systems and the cellular level. Unfortunately, this has not been carried out in regards to the digestive system. However, on the multiple-choice portion of the test, Student 12 gave correct answers to all questions pertaining to the digestive system and glucose. This suggests that she did have at least minimal understanding of the digestive system although she was either unable to integrate this understanding into her drawing and explanation, or that she did not feel the need to include that information here.

The important element in Student 12's post test, is the spontaneous drawing of a flow diagram to help her reason about the question and to explain the causal relationships. Using this flow diagram, it would be possible to interject breakdowns in the system and to conduct diagnostic analysis about both causes and effects.

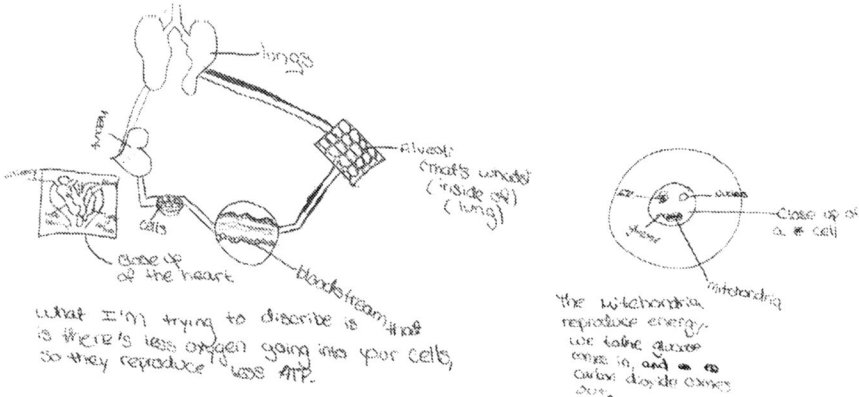

Fig. 13.4 Student12's post instruction model of energy in the human body at the organismic and cellular levels

13.4 Discussion

Quantitative results indicate that the predominately low-income Hispanic middle school students showed significant gains in their understanding of human respiration when they were taught with a curriculum designed on model based learning theory. Students especially showed improvement in their understanding of structure and function of systems and in integration between systems. Many students also improved in making their models dynamic and in a minimal level of causal reasoning. However, analyses of open-ended responses suggest that it was difficult for students to begin with the symptoms in an application problem and work back through a model to determine why the symptoms occur. Some students were able to begin with a change in the system and predict what the symptoms may be, while others could describe the integration and functioning of systems but remained at a naïve level in the application problem. It is possible that there is a continuum students move through in developing their ability to reason with a model. In addition, evidence suggested that students who had developed a dynamic, integrated model were better able to give causal reasons for new phenomena. This suggests that in order to reason with a model in a new situation – apply the model – one may have to first develop a dynamic, integrated, and complex model with which to work.

When students from this inner city trial were compared to students at other schools, their scores were equal to or better to those of the other schools, particularly on the open-ended question. This supports our hypothesis that most students can benefit from using the model based learning approach exemplified in the Energy Curriculum, regardless of ethnicity, language proficiency, and/or socio-economic status. In a formative vein, our in depth analysis of open ended post test items leads us to suggest several further strategies that could be used for fostering reasoning with a model in future implementations. This include, (1) providing more opportunities to practice reasoning with a model provided during instruction; (2) scaffolding of students in working through the model forwards, following the causal progression through to the end; (3) having the teacher present a breakdown in the system such as lack of glucose to the mitochondria, and this causal chain followed to determine what the effect would be and how this effect would be evidenced by symptoms; and (4) having the teacher facilitate the students in working backwards from the symptoms that are evident, using the dynamic model to uncover what is happening. Students need exposure to repeated reasoning with their models even when their model is far from complete. Not only could this improve their ability to use the model in real life applications, but it may also make vivid to the student where the model breaks down and provide impetus for modification of the model.

An important feature of one student's flow diagram presents a starting point for possible instruction of students on how to draw dynamic models that allow them to reason about those models. This particular flow diagram might be used with other students to help them reason through processes and see connections they were unable to see.

This study provides one source of evidence arguing that inner city students are capable of learning complex models in science. In the present case inner city middle school students were able to construct complex visualizable models in biology and engage in model based reasoning with those models. The area where students were less successful in was that of reasoning with their model especially in a diagnostic way. However, this result was no different than in other schools where the research was conducted and valuable information about the use of flow diagrams to promote reasoning with a model were uncovered in the inner city group of students. Future research need to be conducted on how to support students in using pre-made flow diagrams and then gradually initiating their own diagrams.

13.5 Implications for Model-based Teaching and Learning

The ability to construct complex mental models that are functional and structural, and then to use these models is a goal of model-based teaching and learning. Evidence of inner city students' success in accomplishing this goal using a model based curriculum was an important finding of this research. However, even more significant may be the new information gained about student reasoning using flow diagrams. The one student who spontaneously drew a flow diagram was apparently able to use this diagram to describe diagnostic causal relationships. Application of knowledge at this level is a challenging task for students and evidence from over five years of research has suggested that few are able to attain it. It may be that the use of flow diagrams presents a possible inroad into helping students overcome this hurdle. This suggests that future research should focus on strategies to introduce the flow diagram, possibly beginning with teacher generated diagrams. The teacher could then use think-aloud to model use of the flow diagram in diagnosis, followed by scaffolding of student use of the flow diagram. Eventually, it may be that students would be able to generate their own flow diagrams. Any concept that presents causal relationships has the potential for this type of strategy.

References

Clement, J. (2003). Imagistic processes in analogical reasoning: conserving transformations and dual simulations. Paper presented at the American Cognitive Science Society Meeting, Chicago.

Driver, R., Asoko, H., Leach, J., Mortimer, E., & Scott, P. (1994). Constructing scientific knowledge in the classroom. *Educational Researcher, 23*(7), 5–12.

Newton, P., Driver, R., & Osborne, J. (1999). The place of argumentation in the pedagogy of school science. *International Journal of Science Education, 21*(5), 553–576.

Nunez-Oviedo, M. C. (2004). *Teacher-student co-construction processes in biology: Strategies for developing mental models in large group discussions.* Published doctoral dissertation, University of Massachusetts, Amherst.

Nunez, Oviedo, M. C., Rea-Ramirez, M. A., Clement, J., & Else, M. J. (2002). Teacher-student co-construction in middle school life science. Paper presented at ASTE, Charlotte, NC.

Rea-Ramirez, M. A. (1998). *Model of conceptual understanding in human respiration and strategies for instruction* DAI - 9909208, University of Massachusetts, Amherst.

Rea-Ramirez, M. A., Nunez-Oviedo, M. C., Clement, J., & Else, M. J. (2005) *Energy in the Human Body Curriculum.* University of Massachusetts, Amherst.

Rea-Ramirez, M. A. & Else, M. J. (2003). Evidence of model based reasoning in human respiration. *Proceedings of NARST 2003 Conference, PA.*

Chapter 14
Six Levels of Organization for Curriculum Design and Teaching

John Clement

University of Massachusetts

14.1 Introduction

This chapter provides a theoretical perspective on the other chapters in this book by placing the teaching strategies they identify at different levels within a larger organizing framework. Examples of teaching strategies have been drawn from three very diverse contexts: Middle School Biology, High School Mechanics, and High School Electricity. I will examine the possibility that there are teaching techniques that cut across these disciplinary boundaries and age levels. Each of the classrooms studied in the above areas used recently developed model-based curricula that fostered unusually active learning processes. The curricula were designed to develop flexible mental models in students as a key source of understanding. Mental models in areas such as understanding the structure of the lungs or cells, sources and directions of mechanical forces, or causes of current flow in electricity are notoriously difficult for many students to learn. Yet, these lie at the core of conceptual understanding in these areas. Different authors in this book have dealt with teaching techniques at different levels from micro to macro – from individual statements in dialogues to relations between units in a curriculum. This chapter attempts to collect together a number of the identified strategies and to develop an overall theoretical description of teaching techniques at multiple levels.

In order to constrain the task, the focus in this book has been on analyzing *cognitive* strategies, while only touching on metacognitive or motivational strategies, despite their import. We have also concentrated largely on strategies aimed at the construction of *qualitative explanatory models*, rather than at factual knowledge or quantitative models, although examples of connections to quantitative ideas were introduced in Chapters 4, 8, and 9. There was also a focus on strategies for *large group discussion* that involve social interactions; some of these will also apply to small group discussion and individual learning, but there are additional supportive tactics needed there that are not discussed here. I believe that all of the above types of strategies are extremely important, but the hope is that focusing on a few types here has allowed us to gain a foothold in describing those types at a new level of detail within the cognitive core of model based learning.

This book has focused on qualitative explanatory models that were assumed to be at the center of scientific theories being learned. Chapter 2 described explanatory models, such as electrical currents, molecular reactions, or our circulatory system, as providing a description of a hidden, non-observable process that explains how a system works and answers "why" questions about where its observable behavior comes from. This means that the emphasis has been on working causal mechanisms, not just static structure. A common theme in the chapters is that, from the student's point of view, such models can be quite complex, and learning them can involve a series of dozens of conceptual steps.

14.2 General Instructional Techniques Common to the Three Curricula: Co-construction, Evolution

In considering other general techniques that have cut across lessons in different subject areas, we can divide this topic into two main techniques: finding appropriate curriculum Goal Structures describing content targets and sequencing, and finding Teaching Strategies that move students toward those goals.

14.2.1 Common Goal Structures

A distinguishing feature of these model based teaching and learning curricula is their intentional plan to uncover and take into account students' preconceptions, including both useful ideas and misconceptions. An important difficulty with using the current science standards in the USA is that they only specify a target concept or model at best. The specification of the *learning pathway* is missing (Scott, 1992; Clement, 2000;

Niedderer, 2001; Rea-Ramirez, 1998). This can be envisioned in simplest form as a chain starting on the left, with common misconceptions and possible positive preconceptions to build on, and progressing toward the right toward a target model. In between are intermediate models that may be model elements or partial approximations (see Fig. 2.2). Instead of yielding a *logically sequenced* network of true target models in prerequisite order, this yields a *developmentally* sequenced network of partial models in an order that can be traversed by students at a certain age and ability. Learning pathways at the curriculum level were discussed in Chapters 3, 5, and 9 and constitute curriculum goal sequences that take into account students' preconceptions, intermediate models, and appropriate final target models. These can provide a much more fine-grained and focused guide to the teacher concerning what pathway of learning can make sense to the student and lead to deeper conceptual understanding. Pathways that stretch across large topic areas, such as different systems in the human body, set up the important goal of making a curriculum coherent by integrating the student's knowledge into an interconnected framework of ideas.

14.2.2 Common Teaching Strategies

- *Evolution.* In this book, extended case study examples have been provided for a central overall strategy of model evolution that cuts across the three subject areas of biology, chemistry and physics. *Model evolution* contrasts with the simpler approach of model presentation. Teaching via model evolution attempts to make a series of revisions in the student's initial model until the final target model is reached. Figure 14.1 is an abstract version of Fig. 6.3 showing an evolving student model that can be influenced by various observations, analogies, and internal conflicts. As such, it is a way to think about model evolution in any content area.

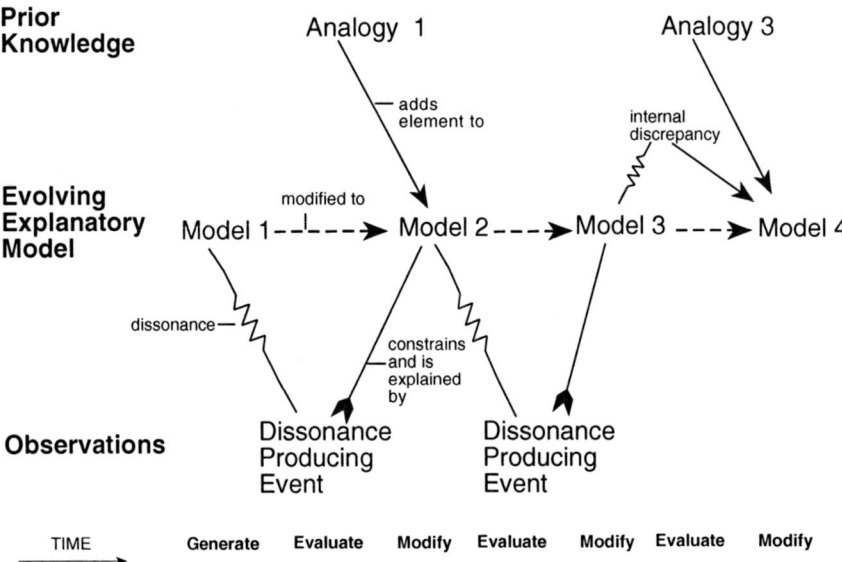

Fig. 14.1 Evolving student model that can be influenced by various observations, analogies and internal conflicts

- *Model criticism and revision processes are part of this approach.* Since these are not emphasized in traditional teaching and learning, it stands to reason that they must be learned as a mode of operating in the classroom, by both the students and the teacher, and this takes time.
- *Co-construction.* It was hypothesized that for classes where the primary goal is learning of conceptual content with understanding, gradual model evolution makes possible *teacher-student co-construction*, with both teacher and students contributing ideas. This is seen as a middle road between lecture and open-ended discovery learning.

This led to the development of a theory for how such an approach has advantages for the learning of explanatory models in science. By starting from students' faulty but interesting ideas, model evolution allows one to: (a) engage students in the fundamental scientific reasoning processes of model evaluation and revision; (b) make revisions in small steps that students can follow and that avoid too much dissonance; (c) deepen understanding by contrasting nature's mechanisms to other, less adequate ones (See especially Chapters 2, 5, 6, 10). One of the most important roles of a partial explanatory model is to serve as a stepping-stone to the next model.

14.2.2.1 Dealing with Both Positive and Negative Aspects of Preconceptions

Scott, et al., (1992) and Clement and Rea-Ramirez (1998) discuss how early approaches to conceptual change teaching tended to emphasize producing dissonance via techniques such as discrepant events, followed, in reaction to this, by approaches that instead built on the student's ideas, often by using analogies. These two opposite sounding strategies can be thought of as represented by the two separate relationships shown in Fig. 14.2. The case studies in this book illustrate how both of these techniques can be combined in the same lessons. The simplest approach to this is to consider the two parts of Fig. 14.2 to be used in sequence in the same lesson, with time going from left to right. In fact, in many cases in this book, the strategy was more complex – something like that shown in Fig. 14.1, with multiple dissonance-producing events and multiple analogies being used to foster a series of model modifications. This is a more complex view of instruction than most curricula advocate. It combines the themes of rationalist and empiricist approaches to learning, and more broadly, the themes of generative and evaluative contributions to explanatory model construction and criticism.

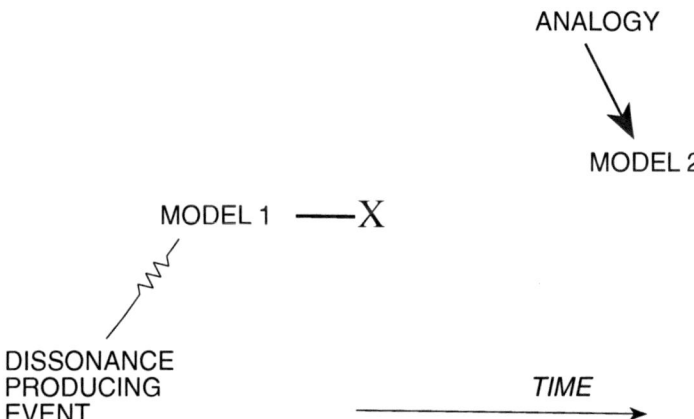

Fig. 14.2 A combination of dissonance producing event followed by a construction producing event such as an analogy

In the old view, there appear to be only two choices: either the students' initial model (preconception) is faulty and should be disconfirmed through dissonance, or the students' analogue model is largely correct as an anchoring intuition and should be appropriated and used in model construction with minor adjustments, as shown in Fig. 14.2. Which is true would seem to dictate whether to use a dissonance strategy or an appropriation strategy. In the present case studies, however, most student models in a topic area are partly correct and partly faulty. The teacher then tries to promote conflict with the faulty pieces while recruiting the positive pieces for model construction, one step at a time, as in Fig. 14.1. Thus, this evolutionary strategy *combines the seemingly "incompatible" strategies of fostering dissonance and appropriating the student's useful preconceptions.* (See especially Chapters 4, 5, 6, 12)

This changes our view of, say, a discrepant event from something that eliminates a preconception to something that can help make it better – from something designed to remove a misconception, to something designed to cast doubt on a particular feature of a partial intermediate model so that it can be improved. Similarly, the idea of using *multiple analogies* changes our view of an analogy from something that creates a final model to something that adds a part to a developing model so that it can be improved (Spiro, et al., 1991). This takes one from the idea that explanatory models *are essentially equivalent to an* analogy to the idea that individual analogies *contribute to the construction of* an explanatory model (see Clement, to appear). Thus, I believe that the model evolution approach to instruction fundamentally changes our view of both analogies and discrepant events (or more broadly, dissonance producing tactics) in a parallel way.

14.2.2.2 Multiple Responsive Strategies

Extending this theme beyond the combining of dissonance-producing strategies with analogies, case studies such as those in Chapter 10 show even more strategies being used in a single lesson, such as requests for explanation, discrepant questions, requests for model evaluation, providing data, requests for supporting and conflicting evidence, providing a model element, or providing demonstrations or animations. Despite the tendency of researchers to study or advocate one favorite type of teaching strategy, these studies reveal a need for the use of *multiple strategies within a single lesson*. Part of this need may be due to the need for several different kinds of conceptual change at different stages within a lesson (Clement, to appear).

To help organize this potentially complicated number of strategies and influences for teachers, another function of a planned learning pathway (e.g., in the central row of Fig. 5.5) is to serve in a curriculum as a *detailed sequence of goals*. Maintaining class discussions, or using student *"voting"* techniques and other ongoing assessments, are ways to give the teacher enough feedback to decide how to keep students in a "reasoning zone" – to decide when to let discussion take its course, when to add more strategies for the present goal, and when to move on to the next goal. Such decisions cannot all be planned in advance and will depend strongly on the particular ideas invented by the students and the level of persistence in preconceptions in that area sensed by the teacher. *This kind of teaching responds to students' ideas and contrasts sharply with teaching that simply uses a lesson plan as a series of topics (i.e., facts) to be "covered" or activities to be completed.*

14.3 Multiple Time Scale Levels of Organization for Model-based Teaching Strategies

Presented in this section are six levels of organization for teaching strategies described in the book (Table 14.1); these strategies range from those that outline the basic shape of an entire curriculum to those that influence the next utterance made by a teacher in class. Complex machines such as DVD players are designed using plans at different levels of systems and subsystems, such as a major components layer, functional subsystems within each component, processes within each subsystem, and so on, with actual circuit diagrams that include the smallest, electronic pieces only at the bottom layer. Similarly, complex computer programs are written in multiple layers of processes and subprocesses. The extremely complex activities of teaching and learning also need multiple layers of organization in order to succeed; yet they are rarely described this way (but see Schoenfeld, 1998).

Each layer in Table 14.1 effectively shows a teaching strategy broken down into substrategies, some of which are in turn broken down at the next lower level. The six levels in Table 14.1 also correspond to a particular time scale, ranging from those strategies operating over months (e.g., sequence of units in a curriculum) to those operating over seconds (e.g., teaching tactics for responding to individual statements in a discussion). Often, in science education, we conduct a study of one particular teaching strategy level, but rarely do we have a chance to show the connections between multiple levels.

Table 14.1 Goals and strategies at different time scale levels for curriculum design and teaching

Level and time schedule	Teaching strategies	Goal structures for student learning outcomes
6-Curriculum Integration Strategies 2–6 months	**Sequencing and Making Connections for Integration** between Major System Models from each Unit	**Integrated Target Structure:** Connections between Units for Integration between Targeted Systems
5-Unit-sized Modeling Strategies 3–15 days	**Major Phases within Each Unit:** Introducing Problems, Building Model Parts, Synthesis, and Model Application	**Top Level Target Models** for Each Unit
4-Lesson Strategies 10–80 Minutes	**Large-scale Model Construction Modes** for Discussion involving multiple models (e.g., Model Competition, Model Combination, Concept Differentiation or Integration, Model Evolution)	**Planned Learning Pathway** Leading to a Target Model for Each Lesson that Explains one or more Target Cases
3-Single Model Element Strategies 0.2–15 Minutes	**Promotion of Individual Model Generation, Evaluation or Modification Processes; or Observation Processes: Small-scale Modes for Class Discussion**	**Model Element** Targeted for Small Conceptual Change or Confirmation; Individual Steps of Conceptual Change from Model M to Model M' in **Implemented Learning Pathway**
2-Individual Cognitive Strategies for Teacher "Moves" in Discussion 5–100 seconds	**Small Cognitive Strategies** (e.g., request for explanation or prediction, request for model evaluation, discrepant question, discrepant event, request or introduce an analogy or bridging analogy, etc.)	Unique Aspects of Student Ideas and Teacher Improvisations Influence Second to Second Acts of Constructive Reasoning on the part of Student as Micro-Contributions to **Implemented Learning Pathway**
1-Dialogical Tactics 1–20 seconds	**Dialogical Tactics** used by a teacher in a single discussion turn for fostering student contributions and sharing them (e.g., reflective toss, indicating respect for ideas, etc.)	**Active Idea Sharing and Social Norms** for Discussion in Science Class

Level 6-*Curriculum Integration Strategies*, refers to strategies for designing the integrated target structure of an entire curriculum. This includes the ordering of the units and how the curriculum might support students developing integrated models that make connections between the units. The example used in the respiration curriculum was the goal of having students be able to connect the pulmonary, digestive, circulatory, and cellular respiration systems by explaining how they all contribute to powering the muscles. Strategies discussed in Chapter 3 include: determining whether the target concept can be achieved within the time frame available, focusing in a disciplined way primarily on only those goals that are essential to achieving the final target concept, determining the form of assessments for measuring integration. A central idea is that determining order of units in terms of ease of learning and prerequisite structure depends on an awareness of key misconceptions and useful preconceptions students enter with.

Level 5-*Unit-Sized Modeling Strategies*, refers to strategies for structuring a unit lasting 3-15 days, (discussed in Chapters 3, 5, 6, and 9). In the respiration curriculum, each unit is written in the following structure (there called a Macro Cycle): Introducing Problems, Building Model Parts, Synthesis, and Model Application (Rea-Ramirez, 1998; see also Driver & Scott, 1996). In a model evolution approach, within the Building Model Parts section, one needs to specify a sequence of top level target models that starts from anticipated preconceptions and leads to the targeted explanatory model for that unit. In addition, choosing evocative and transparent notations is an important strategy for expressing and supporting visual models (discussed in Chapters 5 and 9). Identifying memorable benchmark lessons (Minstrell & Krauss, 2005) that can utilize available local resources for demonstrations, labs, or community issues, so that other lessons can be organized around these, also belongs at this strategy level.

Level 4- *Lesson Strategies*, deal with a planned learning pathway for a single lesson that specifies intermediate models between preconceptions and a target model or subtarget (Chapters 1, 7, 8, 10, 12). Strategies for moving from model to model include Large-scale Model Construction Modes for large group discussion, including Model Competition, Model Combination, concept Differentiation or Integration, and Model Evolution (Chapters 7 and 10). Strategies at this level are distinguished from those at level 3 by the fact that they refer to patterns involving more than one model or more than one model modification. For example, various models of the structure of the throat generated by students in small groups were compared, evaluated and improved in an episode of *model competition* in

Chapter 7. Other strategies deal with supporting imagery, e.g.: "The goal of the experiment is not measurement and confirmation of a principle, but enabling students to run mental simulations that reveal consequences of their existing model and of proposed model modifications." (Chapter 5) Model evolution was the most common strategy at level 4 in this book. It can involve multiple cycles of model generation, evaluation, and modification (GEM cycles) (Chapter 5, 6, 10). Individual processes in this cycle are important strategies at level 3.

Level 3- *Single Model Element Strategies*. Strategies at this level are designed to make or evaluate a single modification in a model and include plans to promote: observations, initial model generation, model element evaluation [disconfirmation, confirmation], or model modification [addition, subtraction, or replacement] (Chapters 4, 7, 8, 10, 11). Modifications at this level correspond most closely to small, individual processes of incremental conceptual change. Part of the model evaluation goal at this level can be that of collecting and finding patterns in observations.

Level 2- *Individual Cognitive Strategies*, are used for individual teacher "moves" in discussion. These can include: requesting oral, written or drawn explanations; requesting or introducing an analogy; introducing an animation, or demonstration; requesting a prediction or example; asking for observations; asking for criticisms or supporting evidence; using discrepant questions or discrepant events. These can be used to implement the processes sought at level 3. For example, a discrepant question can be used to implement the higher order strategy of model element evaluation at level 3, just as the introduction of an analogy can be used to implement the strategy of initial model generation. It should be noted that students may contribute at this level and some of the other levels spontaneously, as in the case of the student who referred to one-way valves in the leg veins as like "lobster traps" (in Chapter 10, which also exhibits many of the above strategies). The list of strategies at this level can be used to expand the idea, represented in Fig. 14.1, of combining the strategies of discrepant events and analogies to contribute to a series of model evaluations and modifications. Many other level 2 strategies can also contribute to such model evolution sequences, highlighting the possibility of the use of multiple strategies within a single lesson. This is a more complex view of conceptual change teaching strategies than is commonly presented (see Clement, to appear).

Level 1- *Dialogical Tactics*. A final level that has not been dealt with much in this book because it is well covered elsewhere is the bottom level.

Readers interested in this level are referred to van Zee and Minstrell (1997) for a discussion of dialogical tactics for setting up classroom norms for fostering and sharing student contributions in general. Chapter 1 begins by discussing strategies at level 1 that are surprisingly hard for a teacher to learn. These start with the much needed strategy of asking more generative questions and deferring judgment on student ideas in order to engage students in active idea sharing and in order to obtain diagnostic information on their existing ideas without suppressing them. This can involve unlearning some very strong habits for many of us.

In sum, the chapters in this book attempt to make contributions to research on teaching that deal with a broad span of strategies from levels 2 through 6 in Table 14.1. The table outlines the way in which curriculum planning, lesson planning, and teaching in science, when taken seriously, is a complex, multilayered process.

14.3.1 Relations Between Levels

In Chapters 7 and 10 Nunez, et al., focused on tracking strategies at levels 3 and 4 in teacher led discussions. They show how strategies at level 3 can be nested within a strategy at level 4 to implement it. Williams and Clement (2006) attempted to unpack and contrast instances of the three sets of strategies at levels 1, 2 and 3 in a case study of electricity instruction; one of their major conclusions is that one can distinguish between strategies at these three levels and connect them to teacher moves in a detailed transcript analysis. They show via diagrams how Cognitive Strategies at level 2 can be used "on top of" Dialogical Tactics at level 1 to foster model evolution processes at the third level.

Comparing level 3 to those above it in the right hand column of Table 14.1, one can distinguish between a "planned learning pathway" specified ahead of time in a lesson plan and a more detailed "implemented learning pathway" that results from the teacher using the plan with real students adaptively. As students introduce unanticipated ideas and details, the implemented pathway is bound to be longer and somewhat different from the planned pathway. Nevertheless, the planned pathway is seen as quite valuable for generating successful implemented pathways.

Conceptual change processes would appear to be affected most directly by strategies at levels 2 through 4 of Table 14.1. For a theoretical overview of types of conceptual change that can be triggered, see Clement (to appear). Some of these go beyond those discussed in this book. For example, di Sessa (1988) introduced the idea of extending the domain of a schema, i. e., changing the conditions of applicability of a schema rather than changing the structure of the schema. Clement (to appear) describes using bridging analogies as a particularly useful strategy (level 2) for fulfilling the goal of extending a model's domain of applicability (level 3). Combining a comprehensive framework for conceptual change types with the present framework for teaching strategy levels may take us a step closer to having an applied theory of conceptual change.

It should be noted that the specifications for the levels in Table 14.1 are simplified generalizations; exceptions to such generalizations certainly exist. For example, analogies can be employed at different levels. They appear to come in different "sizes," and introducing certain complex, large scale analogies can take days, not minutes. Similarly, certain observations may come from a lab that takes much longer than a few minutes. However, the idea of substrategies feeding higher level strategies for each level is a more central point in the table than the exact time scales indicated, which are intended as rough guidelines.

14.4 Comparing Strategies Across Subject Domains: Theoretical Frameworks Regarding GEM Cycles

14.4.1 Similarities

At the center of each step in model evolution is the GEM cycle. In simplest terms this is the cycle of improvement that allows a model to grow better and better as it evolves. It refers to **model generation, evaluation and modification processes.** This was introduced in Chapter 2 as the main driving process for model construction used in the present approaches. Its form is illustrated at the bottom of Fig. 14.1. Historically, these cycles were identified in expert reasoning (Clement, 1989; Nersessian, 1992), leading to the suspicion that they were also central in student learning (but see Driver, 1983).

Table 14.2 shows a comparison of the concept of GEM cycles as used in the three main curricula discussed in this book. It expresses the similarities that constitute a major learning pattern that cuts across the three subject domains and age levels considered. Supporting GEM cycles is proposed here as a general teaching strategy that all of these domains share in common.

In order to construct Table 14.2, I am interpreting Samia Khan's Chapter 4 as documenting GEM cycles in a tutoring case study. Although a major tool of the teacher is to ask for predictions, he has designed these

Table 14.2 Comparing findings: theoretical frameworks regarding GEM cycles

Phases	College chemistry framework (Guided discovery approach)	Middle school biology framework	High school electricity framework
G Generate Model	Teacher challenges the students to make a prediction from their initial model or from a data table of observations	Detecting students' ideas – often via teacher's request for an explanation; If very few ideas present, teacher may introduce an initial analogy	Teacher or Manual asks students to make prediction for circuit (this encourages them to generate a model if they do not have one)
E Evaluate Model	Students are introduced to new anomalous data violating their prediction and asked to evaluate their generalization	Teacher attempts to generate dissonance: often via discrepant questions but sometimes by asking students for idea evaluation	Students observe circuit behavior (often a discrepant event, but sometimes confirmatory); Students encouraged to map implications to their current model to evaluate it
M Modify Model (Student Contribution)	Students are prompted to modify their generalization about behaviors; Students are asked to explain new findings	Students asked to modify (or sometimes disconfirm) their present model	If faulty, students modify model in discussion in order to account for observations
M Modify Model (Teacher Contribution)	Teacher aids modification where necessary with positive feedback, hints, additional information, or explanations	Teacher aids modification where necessary with analogy, hints, pictures, animations, or explanations	Teacher aids modification where necessary with analogy, hints, notations, or explanations

questions to be ones that get students to construct and articulate general properties of molecular models in order to make the prediction – for example to predict the effect of molecular weight on boiling point. Thus he is implicitly asking them to generate a model. Comparing their predictions to simulation data then comprises a way to evaluate the model. Khan documents a series of model modifications in such a cycle.

The strategy of fostering a component of the GEM cycle appears in Table 14.3 at level 3. This can be supported most directly by teaching moves that simply ask students to perform the three processes, e.g. "Can you generate an explanation (by generating a model) for that observation? Can you evaluate whether that model makes sense? Is it supported? Can you modify it?" However, each of these can in turn be supported by the two levels of strategies below, at levels one and two. This kind of scaffolding via questioning is fairly directive on the part of the teacher; despite this, it can lead to ideas that are largely student generated.

Another common pattern was described as students evaluating their explanatory models by running them as mental simulations (Chapters 11, 5, 4). This activity is seen as what produces predictions from the models during these cycles. When viewed as the use of imagery, this is coherent with the fact that each curriculum recommends using student generated drawings during learning, presumably to support such imagery. This is particularly apparent in Chapters 4, 7, and 12.

14.4.2 Differences

One difference between the curriculum areas reflected in Table 14.2 is that the chemistry cases often combined an empirical cycle (or "quasi empirical", using a simulation) followed by an explanatory model cycle, as did the electricity curriculum. On the other hand, the biology curriculum often used rational reasoning from cases more frequently than it did experiments. This demonstrates that a GEM cycle can be present in either context.

In summary, the GEM cycles described in Table 14.2 appear to be a central common teaching and learning strategy in all three curriculum contexts. The grain size of the time scale for this strategy is seen as being at level 3 in Table 14.1.

14.5 Constructive Reasoning and Argumentation Reasoning

A good deal of valuable prior work on student discussion has focused on student *argumentation* processes, e.g., Duschl and Osborne (2002), Clark and Sampson (2005). This work builds on Toulmin's (1958) categories for scientific reasoning. Some of the processes documented in this book can be thought of as augmentation, especially those categorized as model competition in Chapter 7. But we have also focused on a number of other processes within GEM cycles that we consider central that do not seem to fit into the category of argumentation. Essentially, argumentation, as discussed by authors such as Duschl and Osborne (2001), involves *evaluating* alternative theories or models. I believe it is also important to focus on *constructing* theories by *generating and modifying* models.

Certainly, there are many more elaborate theories of the scientific inquiry process, but the GEM cycle appears to be the least common denominator, or minimum common process, within them. When an investigation includes empirical work, we may refer to an OGEM cycle to include the contribution of observations prior to model generation as well as during model evaluation (Williams & Clement, 2006). I use the term *constructive reasoning* to refer to the many types of individual reasoning operations that can support the OGEM cycle. This includes argumentative reasoning in which students debate the merits of one or more models during the evaluation phase. However, it also includes more generative types of reasoning, such as inductions, abductions, and analogies, in the more generative phases during which observation patterns are formed and explanatory models are generated and modified.

Thus Level 3 in Table 14.1 includes the processes of model generation (including visualization of model dynamics) and model modification. These processes were especially highlighted in Chapters 1, 4, 6, 8, and 10, although the theme of model modification really does pervade the entire book. These are student-learning processes that are central to building models. They lie outside the domain of argumentation, basically because they concern processes for building up a model rather than arguments for evaluating the validity of a model.

One should ask whether this comparison can be made, since argumentation tends to refer, in much of the literature, to *student reasoning* rather than to teaching strategies (but see Osborne, et al. 2001). Table 14.1, on the other hand, is concerned with *teaching strategies*. However, these teaching strategies are all directly aimed at supporting certain student reasoning and learning processes, and so they are closely linked to them. Level 2 in Table 14.1 includes specific teaching strategies such as requests for explanation, requests for modification, a request for or introduction of an analogy, or engaging with animated simulations. These teaching strategies are aimed at promoting model construction via types of constructive reasoning that are often not argumentative. The difference between the approach of this book and the argumentation approach is understandable, because the focus in this book is on content learning via model construction, and I believe the approaches are complementary.

14.6 Conclusion

In sum, this chapter integrates findings from the studies in this book of instructional strategies for attaining deeper levels of conceptual understanding in science. These studies have all been concerned with the problem of how to get students actively engaged, not just with hands on activity, but with the cognitive activity of generating, evaluating, and modifying mental models. This activity is seen as lying at the core of learning science with understanding. Some general teaching strategies were found that appear to cut across science curricula and grade levels: model evolution, fostering GEM cycles, co-construction, building on positive preconceptions, etc. This is of interest, because there has been an insufficient exchange of information between the sciences on teaching strategies. In particular, examples of co-construction presented, where both the student and the teacher contribute ideas for model construction, illustrate that this is not an either-or choice in education. New ways of describing co-construction, such as the diagrams in Chapter 10 showing both teacher and student contributions, illustrate the sense in which the theory of model-based co-construction developed in this book attempts to integrate social and cognitive perspectives for explaining science instruction.

Secondly, it is possible to organize these and other strategies into six time scale levels, shown in Table 14.1. There is insufficient awareness in education of the multiple levels of model construction strategies that can be used in teaching. Viewing the strategies in this way allows one to see how lower level strategies act in support of higher level strategies. Chapters are keyed to the different levels and contribute support to the outline of a theory of instruction that is broader than normally discussed. We suspect that instructional design must include all six levels to be optimally effective for teaching for meaningful conceptual change leading to integrated knowledge that can be applied flexibly. Thinking about curriculum at different time-scale planning levels is not, in itself, a new idea, but there has been insufficient work that deals with multiple levels and that is grounded in cases studies of actual curricula and of actual classroom interactions. Issues such as the distinction between dialogical and cognitive strategies, the importance and difficulty of managing an agenda in large group discussion, and the pervasive role of model construction cycles of evaluation and revision have been understudied in the science education literature. By separating strategies according to time scale levels, we create the potential to help teachers sort out different levels for planning and for structuring discussion.

The intent behind the approaches described in this book was that students who participate actively in developing models via GEM cycles will have a deeper understanding of and appreciation for the remarkable structures and mechanisms of nature as well as for the creative but critical process of science itself. Starting in Chapter 1, it was argued that one can foster a significant degree of inventive, student initiated idea generation and evaluation while still retaining a significant degree of control over agenda setting for content goals. Thus we hope this book will help educators in their quest to find stimulating and successful methods that lie between pure discovery and lecture based approaches to science instruction.

14.7 References

Clark, D. B., & Sampson, V. D. (2005). Analyzing the quality of argumentation supported by personally seeded discussions. *Proceedings of the 2005 Conference on Computer Support for Collaborative Learning* (pp. 76–85), International Society of the Learning Sciences. May 30 – June 04, 2005, Taipei, Taiwan.

Clement, J. (1989). Learning via model construction and criticism: Protocol evidence on sources of creativity in science. In J. Glover, R. Ronning, & C. Reynolds (Eds.), *Handbook of creativity: Assessment, theory and research.* NY: Plenum, 341–381.

Clement, J. (2000) Model based learning as a key research area for science education. *International Journal of Science Education*, 22(9), 1041–1053.

Clement, J. (to appear). The role of explanatory models in teaching for conceptual change. In S. Vosniadou (Ed.), *Handbook of research on conceptual change.* Mahwah, NJ: Lawrence Erlbaum.

Clement, J., & Rea-Ramirez, M. (1998). The role of dissonance in conceptual change, *Proceedings of National Association for Research in Science Teaching.*

Cobb, P. (1988). The tension between theories of learning and instruction in mathematics education. *Educational Psychologist, 23*(2), 87–103.

di Sessa, A. A. (1988). Knowledge in pieces. In G. Forman & P. B. Pufall (Eds.), *Constructivism in the computer age.* Hillsdale, NJ: Erlbaum Associates.

Driver, R. (1983). *The pupil as a scientist?* Milton Keynes: Open University Press.

Driver, R., & Scott, P. (1996). Curriculum development as research: A constructivist approach to science curriculum development and teaching. In D. F. Treagust, R. Duit, & B. J. Fraser (Eds.), Improving teaching and learning in science and mathematics (pp. 94–108). New York: Teachers College Press.

Duschl, R. A., & Osborne, J. (2002). Supporting and promoting argumentation discourse in science education. *Studies in Science Education,* 2002; *38,* 39–72.

Minstrell, J., & Krauss, P. (2005). Guided inquiry in the science classroom. In M. S. Donovan & J. D. Bransford (Eds.), *How students learn: Science in the classroom* (pp. 475–514). Washington: National Academies Press.

Nersessian, N. J. (1992). How do scientists think? Capturing the dynamics of conceptual change in science. In R. N. Giere (Ed.), *Cognitive models of science* (Vol. 15, pp. 3–44). Minneapolis: University of Minnesota Press.

Niedderer, H. (2001). Physics learning as cognitive development. In R. H. Evans, A. M. Andersen, & H. Sørensen (Eds.), *Bridging research methodology and research aims*. The Danish University of Education. (ISBN: 87-7701-875-3), (pp. 397–414). http://didaktik.physik.uni-bremen.de/niedderer/personal.pages/niedderer/Pubs.html#lpipt

Osborne, J. F., Erduran, S., Simon, S., & Monk, M. (2001). Enhancing the quality of argument in school science. *School Science Review, 82* (301), 63–70.

Rea-Ramirez, M. A. (1998). Models of conceptual understanding in human respiration and strategies for instruction. DAI – 9909208, University of Massachusetts, Amherst.

Schoenfeld, A. H. (1998). Toward a theory of teaching-in-context. *Issues in Education, 4*(1), 1–93.

Scott, P. H., Asoko, H. M., & Driver, R. (1992). Teaching for conceptual change: A review of strategies. In R. Duit, F. Goldberg, & H. Niedderer (Eds.), *Research in physics learning: Theoretical issues and empirical studies* (pp. 310–329). Kiel: IPN.

Scott, P. H. (1992). Conceptual pathways in learning science: A case study of the development of one student's ideas relating to the structure of matter. In R. Duit, F. Goldberg, & H. Niedderer (Eds.), *Research in physics learning: Theoretical issues and empirical studies* (pp. 203–224). Kiel: IPN.

Spiro, R. J., Feltovich, P. J., Coulson, R. I., & Anderson, D. K. (1991). Multiple analogies for complex concepts: Antidotes for analogy-induced misconception in advanced knowledge acquisition. In S. Vosniadou & A. Ortony (Eds.), *Similarity and analogical reasoning*. Cambridge, UK: Cambridge University Press.

Toulmin, S. (1958). *The uses of argument.* Cambridge, UK: Cambridge University Press.

van Zee, E. H., & Minstrell, J. (1997). Reflective discourse: Developing shared understandings in a high school physics classroom. *International Journal of Science Education, 19*, 209–228.

Williams, E. G., & Clement, J. (April 2006). Teacher moves during large-group discussions of electricity concepts: Identifying supports for model-based learning. *Proceedings of the NARST annual Meeting* – San Francisco, CA.

Author Index

Anderson, C. W., 45, 46
Anderson, D. K., 216
Anderson, R. C., 27
Ascher, R. C., 47
Asoko, H. M., 31, 32, 233
Atwood, Virginia A., 196

Bagno, E., 169
Bean, T. W., 196
Beichner, R. J., 98
Billett, S., 32
Bishop, B. A., 45, 46
Black, J., 31
Black, T., 30
Boulter, C. J., 12, 28, 31
Brewer, W. F., 33, 208
Brown, A. L., 26
Brown, D. E., 12, 26, 27, 29, 31, 99, 216
Brualdi, A., 196
Buckley, B. C., 12, 31
Bulgren, J. A., 32, 216, 222

Camp, C., 45, 112
Campione, J., 26
Catambone, R., 30
Champagne, A. B., 208
Chi, M. T. H., 31
Chinn, C. A., 33, 208
Chiu, M. T. H., 31, 46
Chou, C.-C., 31
Chuska, K., 196
Clark, D. B., 268
Clement, J., 3, 11–14, 19, 26, 29–37, 81, 83, 98, 99, 104, 108, 111–114, 118, 170, 174, 175, 179, 183, 190, 196, 198, 216, 221, 234, 242, 256, 259, 260, 264–266, 269
Closset, J.-L., 98
Cobern, W. W., 26
Cohen, R., 98, 108
Collins, A., 30
Corey, V., 196
Cosgrove, M., 208
Cotton, K., 196, 198
Coulson, R. J., 216
Craig, D. L., 30
Curtis, R. V., 226

Dagher, Z., 215
Darden, L., 28, 190
Davies, J., 30
Davis, J., 211
Deshler, D. D., 32
di Sessa, A. A., 265
Doster, E., 112
Dreyfus, A., 26
Driver, R. H., 31, 32, 233, 263, 266
Duit, R., 31, 98, 198, 222
Dunbar, K., 30
Dupin, J. J., 216
Duschl, R. A., 174, 268, 269

Eliovitch, R., 26
Ellis, K., 196
Else, M. J., 3, 11, 32, 36, 175, 183, 196, 216, 226, 234
Engelhardt, P. V., 98
Ertmer, P. A., 216
Eylon, B. S., 98, 108, 169

Feltovich, P. J., 216
Forbus, K. D., 221

Francoeur, E., 59, 60
Frederiksen, J. R., 31, 114
Fredette, N., 98

Gall, M., 196
Ganiel, U., 98, 108
Gellert, E., 47
Gentner, D., 30, 221, 225
Gertzog, W. A., 24
Gibson, H., 36, 198
Gilbert, J. K., 12, 28, 31
Glaser, B. G., 31, 174
Glaser, R., 31, 174
Glynn, S. M., 31, 112, 198, 216, 222, 223
Gobert, J. D., 31
Goldberg, F., 36, 45, 83, 108
Gooding, J. N., 196
Goswami, U., 225
Gould, S. J., 28
Gruber, H., 28
Gunstone, R. F., 148, 149, 208

Hammer, D., 12
Harmon-Jones, E., 33
Harrison, A. G., 31, 222, 226
Hassard, J., 196
Hatano, G., 27, 175, 215, 221
Hawkins, C., 112
Hennessey, G. M., 31, 114
Hesse, M., 29
Hewson, M., 24
Hewson, P. W., 24
Hogan, K., 31, 32
Holowchak, M., 37, 174, 233

Inagaki, K., 175, 215, 221
Iwasyk, M., 196

Johnson, S. K., 31
Johnson-Laird, P. N., 30, 196
Johsua, S., 216
Jungwirth, E., 26

Kaplan, D., 31
Kelly, A. E., 174

Keselman, A., 31
Khan, S., 60, 266, 268
Klopfer, L. E., 208
Krauss, P., 263
Kuhn, D., 31
Kuhn, T. S., 24, 25, 28
Kurose, A., 196
Kurz-Milcke, E., 30

Leach, J., 32, 233
Leander, K. M., 27
Leeuw, N. d., 31, 46
Lenz, B. K., 32
Leone, T. J., 139
Leven, T., 196
Liu, C.-J., 31
Lochhead, J., 98
Long, R., 196

Maxwell, J. C., 28, 170
McCroskery, 196
Mills, J., 33
Minstrell, J., 12, 118, 137, 196, 263, 265
Mintzes, J. J., 45, 46
Mortimer, E., 32, 233

Nagy, M. H., 47
Nastasi, B. K., 31
Nersessian, N. J., 27, 28, 30, 34, 35, 170, 190, 266
Newby, T. J., 216, 225, 228
Newton, P., 24, 233
Nichols, K., 112
Niedderer, H., 36, 45, 83, 108, 257
Novick, N., 208
Nunez, Oviedo, M. C., 3, 11, 17, 32, 36, 37, 118, 174, 179, 183, 196–198, 216, 221, 234, 265
Nussbaum, J., 208

Osborne, J. F., 174, 233, 268, 269
Osborne, R., 118, 208

Palincsar, A. S., 26
Perkins, D. N., 27

Porter, C. S., 47
Posner, G. J., 24
Pressley, M., 31
Preub, A., 31

Raduta, C., 169
Raghavan, K., 31
Rea-Ramirez, M. A., 3, 11, 31–33, 36, 45, 57, 118, 174, 175, 179, 183, 189, 196–198, 210, 216–218, 233–235, 257, 259, 263
Reiser, B. J., 139
Resnick, L. B., 37, 174, 233
Roth, K. J., 45, 46

Salmon, M., 37, 174, 233
Sampson, V. D., 269
Sanders, M., 45, 46
Sandoval, W. A., 139
Schnotz, W., 31
Schoenfeld, A. H., 261
Schumaker, J. B., 32
Schwartz, D. L., 30
Schweber, S., 28
Scott, P. H., 31–33, 179, 189, 233, 256, 259, 263
Segal, J. W., 27
Seitz, C. M., 174
Shipstone, D. et al., 98
Simpson, D., 196
Slotta, J. D., 31
Smith, B., 139
Songer, C. J., 45, 46
Spiro, R. J., 112, 216, 260
Stahl, R. J., 196
Stavy, R., 26, 226

Steinberg, M. S., 7, 11, 13, 14, 31, 33, 36, 79, 81, 83, 98, 108, 113, 114, 151, 154, 166, 170, 179
Steinmuller, F., 139
Stepich, D. A., 216
Stevens, A. L., 30
Stewart, J., 31
Stimpson, V., 196
Strauss, A. I., 174
Strike, K. A., 24
Swift, Schell, P. R., 196

Tabak, I., 139
Tait, C. D., 47
Takahashi, T., 216
Thagard, P., 24
Thompson, C., 199
Tobin, K., 196
Toulmin, S., 174, 269
Trafton, J. G., 30
Treagust, D. F., 31
Trickett, S., 30
Tweney, R. D., 28

van Zee, E. H., 12, 137, 196, 265
Voss, J. F., 27
Vygotsky, L. S., 19, 25

Wainwright, C. L., 31, 79
Wathen, S. H., 37, 174, 233
White, B. Y., 31, 114
White, R., 148, 149
Wild, J., 196
Williams, E. G., 114, 265, 269
Wittrock, M., 118

Zietsman, A., 12, 29, 31, 36

Subject Index

Accretion, 83, 198
Accretion Mode, 117, 133, 137, 174, 178, 187
Alternative conceptions, 12, 24, 46, 115, 202, 246
Analogies, 5, 8, 30, 46, 54, 81, 96, 103, 108, 109, 112, 114, 154, 168, 187, 215–229
Anchor, 81
Anchoring, 80

Base, 30

Causal agent, 83
Causal reasoning, 236, 247
Causal relationships, 250
Chemistry, 59
Cloud model, 147
Co-construction, 3, 7, 37, 117, 178–191, 234, 258
Co-construction modes, 117, 120, 121
Cognitive model construction, 12
Competition mode, 117, 118, 132, 137
Complex concepts, 49, 153
Complex conceptual model, 49
Complex mental models, 173–191
Conceptual change, 46, 80, 262, 264, 270
Conceptual change theory, 24
Conceptual understanding, 234
Confirmation Mode, 177, 184, 187, 190
Constructive modeling, 34
Criticism and revision cycles, 56, 210

Demonstration activity, 184
Disconfirmation, 190
Disconfirmation Mode, 133, 137, 174, 187, 190
Discovery learning, 26
Discrepant events, 50, 80, 110, 112–114, 264
Discrepant questioning, 8, 195
Discrepant questions, 14, 21, 264
Disequilibrium, 24
Dissatisfaction, 33, 80, 182, 190, 198, 200, 208
Dissonance, 14, 33, 50, 119, 137
Dissonance-producing event, 175
Distant action models, 154
Drawing, 132, 133
Dynamic imagistic model, 166

Element Confirmation, 184, 185
Element Confirmation Mode, 184
Element Disconfirmation, 173
Elements, 226
Energy and the Human Body, 234
Evaluating a model, 191
Evolution, 257
Evolution Mode, 174, 179, 189, 190
Evolving mosaic mixture, 19
Explanation, 30
Explanatory model, 29, 112, 263
Extrinsic dissatisfaction, 182

Formal reasoning, 35

GEM cycles, 37, 190
Generating a model, 191
Generative exploratory studies, 174
Guided inquiry, 15

Subject Index

Hands-on experiments, 80
Hands-on learning activities, 46

Imageable concrete analogy, 168
Imagery, 268
Imagistic reasoning, 151
Inquiry, 37
Intermediate model elements, 210
Intermediate models, 19, 50, 54
Intermediate target, 49

Learning pathways, 34, 45, 47, 52, 55, 68, 79, 189, 256, 260, 263, 265

Mapping, 217
Mathematical modeling, 96
Mental model construction, 195, 197, 198
Mental modeling theory, 27, 28
Mental models, 30, 52, 61, 117, 125, 216
Mental representations, 60
Mental simulations, 30
Misconceptions, 18, 24, 36, 209
Model, 174
Model-based co-construction, 23
Model based learning, 23
Model based reasoning theory, 234
Model based teaching and learning, 252, 253
Model building, 80, 234
Model construction, 19, 23, 31, 50, 61, 109, 179
Model element change, 200, 201
Model element modification, 203
Model elements, 11, 120, 124, 210, 216, 234, 257
Model evolution, 33–35, 111, 114, 258
Model generation mode, 176, 177
Modeling cycle, 105
Model modification, 269
Model modification mode, 184
Modification mode, 176, 189
Mosaic of student ideas, 17

Naive conceptions, 47
National standards, 47, 56
Near vs far, 225
Negative feedback, 119

Optimum integrated target model, 48, 49

Piagetian theory, 24
Planetary model, 145
POE, 147–150
Preconceptions, 197, 270, 261, 263
Prediction, 64
Prediction-making activities, 63
Predict-Observe-Explain, 148

Reasoning, 60
Reasoning with a model, 244, 252
Runnability, 168, 169
Runnable models, 98
Running the model, 202

Science standards, 36
Scientific model, 179, 210
Social learning theories, 25, 26
Source analogies, 107
Student-active, 218
Student directed, 15
Student generated, 21, 268
Student-generated model elements, 234
Student mental model development, 54
Students, 234
Student-teacher co-construction, 32
Supporting questions, 198

Target, 30, 51, 52
Target concepts, 46, 47, 236
Target models, 11, 45, 49–52, 228
Teacher directed, 182
Teacher generated, 182
Teacher-student co-construction, 77, 182
Teaching by telling, 61

Think aloud, 104
Thought experiment, 176, 184, 199
Transfer, 215

Undergraduate, 60

Visual analogy, 187
Visual imagery, 228
Visualizable models, 20

What if scenarios, 139, 148–150

Printed in the United States
127563LV00001B/92/A